Introductory Mathematics

Introductory Mathematics

Kacey Conley

Willford Press
www.willfordpress.com

Published by Willford Press,
118-35 Queens Blvd., Suite 400,
Forest Hills, NY 11375, USA

ISBN: 978-1-64728-350-6

Cataloging-in-Publication Data

Introductory mathematics / Kacey Conley.
p. cm.
Includes bibliographical references and index.
ISBN 978-1-64728-350-6
1. Mathematics. I. Conley, Kacey.
QA36 .I58 2022
510--dc23

For information on all Willford Press publications
visit our website at www.willfordpress.com

Willford Press

Contents

Preface

Mathematics is a discipline which includes the study of various topics such as quantity, space, change and structure. It is classified into two primary areas- pure mathematics and applied mathematics. Pure mathematics studies only the concepts of mathematics. It does not delve into any applications of such concepts in diverse fields such as computer science, engineering and business. Some of the important areas of mathematics are algebra, number theory, geometry, mathematical analysis, arithmetic, etc. It plays an important role in many fields such as natural science, engineering, social sciences, medicine and finance. Advancements in applied mathematics resulted in the development of new mathematical disciplines such as statistics and game theory. This textbook elucidates the fundamental concepts of mathematics. It presents this complex subject in the most comprehensible and easy to understand language. This book will serve as a valuable source of reference for those interested in mathematics.

A short introduction to every chapter is written below to provide an overview of the content of the book:

Chapter 1 - Mathematics is a domain that deals with the study of topics such as quantity, change, space and structure. Pure mathematics and applied mathematics are the two primary areas that fall under this discipline. This is an introductory chapter which will introduce briefly all the significant aspects of mathematics; **Chapter 2 -** Algebra is a significant subset of mathematics that deals with the study of mathematical symbols and the rules for manipulating these symbols. It includes elementary equation solving as well as the study of abstractions such as fields, groups and rings. Elementary algebra and abstract algebra are the two main types of algebra. This chapter has been carefully written to provide an easy understanding of the varied facets of algebra; **Chapter 3 -** The branch of mathematics that is concerned with the questions of size, relative position of figures, shape and the properties of space is known as geometry. Euclidean geometry, Differential geometry, Algebraic geometry and Solid geometry are the various types of geometry. The topics elaborated in this chapter will help in gaining a better perspective about the various types of geometry; **Chapter 4 -** Trigonometry is a mathematical branch that focuses on the study of relationships between the angles and side lengths of a triangle. It is used in various fields such as mechanical engineering, electrical engineering, cartography, computer graphics, etc. This chapter discusses in detail the theories and methodologies related to trigonometry; **Chapter 5 -** The mathematical study of continuous change is known as calculus. Differential calculus and integral calculus are the two main branches that fall under this domain. The convergence of infinite sequences and series with respect to a well-define limit are the common foundations of both branches. All the diverse principles of calculus have been carefully analysed in this chapter; **Chapter 6 -** A mathematical equation that relates some function with its derivatives is known as a differential equation. Ordinary differential equations, partial differential equation and non-linear differential equation are the common types of differential equations. The diverse applications of differential equation in the current scenario have been thoroughly discussed in this chapter; **Chapter 7 -** Probability is a measure that quantifies the likelihood of events that might occur. Statistics is a subset of mathematics which deals with data collection, analysis, organization, interpretation and presentation. The chapter closely examines the key concepts of probability and statistics to provide an extensive understanding of mathematics.

I extend my sincere thanks to the publisher for considering me worthy of this task. Finally, I thank my family for being a source of support and help.

Kacey Conley

1

Introduction to Mathematics

Mathematics is a domain that deals with the study of topics such as quantity, structure, space and change. Pure mathematics and applied mathematics are the two primary areas that fall under this discipline. This is an introductory chapter which will introduce briefly all the significant aspects of mathematics.

Mathematics is the science that deals with the logic of shape, quantity and arrangement. It constitutes a body of established facts, achieved by a reliable method, verified by practice, and agreed on by a consensus of qualified experts. But its subject matter is not visible or ponderable, not empirical; its subject matter is ideas, concepts, which exist only in the shared consciousness of human beings. Thus it is both a science and "humanity." It is about mental objects with reproducible properties.

For example, "the triangle" in Euclidean geometry, or the counting numbers 1, 2, 3, 4, in arithmetic are concepts which we can communicate, and which, as we can verify, keep their properties as they are communicated. These concepts are reproducible, they possess certain rigidity, a reliability and consistency, and so they permit conclusive, irresistible reasoning—which is "proof."

"Proof," not in the formal or formalized sense, but in the sense in which mathematicians mean proof—conclusive demonstrations that compel agreement by all who understand the concepts involved. Abstract concepts subject to such conclusive reasoning or proof are called mathematical concepts.

Mathematics is the subject where answers can definitely be marked right or wrong, either in the classroom or at the research level. Mathematics is the subject where statements are capable in principle of being proved or disproved, and where proof or dis proof bring unanimous agreement by all qualified experts.

Reasoning about mental objects (concepts, ideas) that compels assent (on the part of everyone who understands the concepts involved) is "mathematical". This is what is meant by "mathematical certainty".

Certainly mathematics itself isn't the only place where conclusive reasoning occurs! Rigorous reasoning can occur anywhere-in law, in textual analysis of literature, and in ordinary daily life apart from academics. Historians can use unimpeachable reasoning to establish a sequence of events, or to refute anachronistic claims. But although historical dates are subject to rigorous reasoning, they are not mathematical objects, because they are tied to specific places and persons. Information about them comes, ultimately, from someone's visual or auditory perceptions.

Mathematical conclusions are decisive. Just as physical or chemical knowledge can be independently verified by any competent experimenter, an algebraic or geometric proof can be checked and recognized as a proof by any competent algebraist or geometer. There has been one famous disagreement about valid mathematical proofs, Luitjens Brouwer and Errett Bishop rejected "proof by contradiction." That disagreement resulted in the development of a variant, "intuitionistic" or "constructivist" mathematics. Intuitionistic or constructivist mathematics makes a stricter demand on what is a "rigorous proof." Knowing how to recognize and accept a "rigorous proof" is the condition for membership in the community of mathematicians, whether the usual "classical" or the minority "constructivist" version.

Other, hitherto unthought-of kinds of mathematical behavior will yet arise. A definition of mathematics should accept the yet-to-be-created new mathematical subjects that are sure to arise in coming decades, not to say centuries. How will we identify such hitherto unseen behavior as mathematical? How has it been decided in the past, that some new branch of study is not just "mathematical" (containing some mathematical features), but really *mathematics*—requiring to be included within mathematics itself.

One famous example was probability gambling or betting. Fermat and Pascal demonstrated "rigorous" (irrefutable, compelling) conclusions about some games of chance. Therefore their work was mathematical, even though it was outside the bounds of mathematics as previously understood. Subsequent work of Bernoulli, De Moivre, Laplace and Chebychev was mathematics, for the same reason. Ultimately Kolmogorov axiomatized probability in the context of abstract measure theory. In doing so he was axiomatizing an already existing, ancient branch of mathematics.

A more recent example is set theory. Infinite sets were not part of mathematics before Georg Cantor explicitly based them on the notion of one-to-one correspondence. On that basis, he was able to make compelling arguments, and then set theory (with some resistance) became a mathematical subject.

Since Aristotle, formal logic has helped to clarify mathematical reasoning, and rigorous argument in general. It draws conclusions on the basis of the logical form of statements—their "syntax." But most mathematical argument is based more on the *content* of mathematical statements than on their logical form. It is done without referring to the rules of formal logic, even without awareness of them. In the process of actively discovering or creating mathematics, logicians and other mathematicians reason by analogy, by trial and error, or by any other kind of guessing or experimentation that might be helpful. In fact, formal logic itself is well-established as a *part of* mathematics.

As such, it is subject to conclusive reasoning that is informal, like any other part of mathematics. Logicians reason informally in proving theorems about formal logic.

Numbers

Numbers are strings of digits used to indicate magnitude. They measure size - how big or small a quantity is. In mathematics there are several types of numbers, but they fall into two main classes, the counting numbers, and scalars.

Counting Numbers and Natural Numbers

These are used to count the number of objects. They are positive whole numbers and have no fractional parts. For example 12 cars, 45 students, 3 houses.

Scalars

These are numbers used to measure some quantity to any desired degree of accuracy. For example a building height is 12.388 meters, or speed of an aircraft is 810.31 kilometers per hour. They can have decimal places or fractional parts. Within this category there are several types of number:

Real Numbers

Real numbers are those that can be positive, negative or zero, and can have decimal places or fractional parts. They are the most common numbers used in measuring quantities. Example 31.88 centimeters. They usually have units.

Integers

Integers whole numbers that can be positive, negative or zero, but have no decimal places or fractional parts. They are like the counting numbers but can be negative.

Positive and Negative Numbers

Positive numbers are those which are considered to be greater than zero. A large positive number is larger than a smaller one, for example +12 is larger than +2.

Negative numbers are those considered to be less than zero. They can be thought of as a debt or deficit. For example, if your wallet is empty and you owe someone $12, then you can think of your wallet as having negative $12. In a way you have less than zero dollars.

Rational and Irrational Numbers

Rational numbers are those that can be written as the ratio of two integers.

The word 'rational' comes from 'ratio'. For example the number 0.5 is rational because it can be written as the ratio ½.

Irrational numbers are those that are not rational, that is those that cannot be written as the ratio of two integers.

Imaginary Numbers

Imaginary numbers are those needed to find the square root of negative numbers, which would not normally be possible. So for example the square root of -16 would be written 4i, where i is the symbol for the square root of negative one.

Complex Numbers

Recall that real numbers are those that lie on a number line. Complex numbers extend this idea to numbers that lie on a two dimensional flat plane. Complex numbers have two components called the real and imaginary parts.

Prime Numbers and Composite Numbers

A prime number is an integer that has no factors, other than one and itself. In other words it can be divided only by one and the number itself. 17 is a prime number. 16 is not because it can be divided by 2, 4 and 8.

A composite number is one that is not prime. It does have factors, and so is the opposite of a prime number.

Number Notation

There are various ways that numbers can be written or diagrammed:

The Number Line

A number line is a graphical way to visualize numbers by placing them on a straight line, usually with zero in the middle, positive numbers to the right and negative numbers to the left.

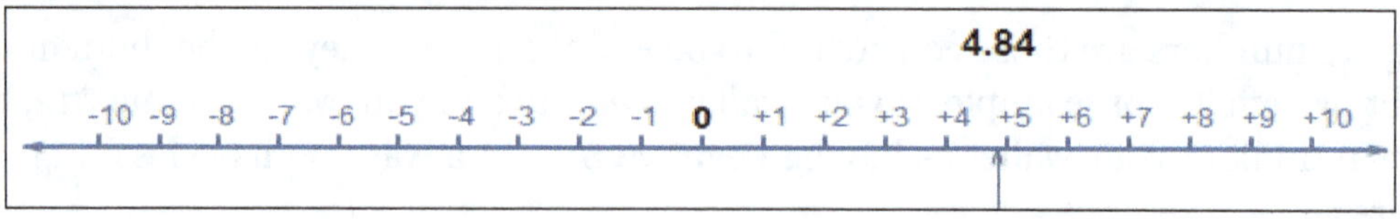

Decimal Notation

The most common way to represent real numbers. A string of digits and a decimal point (dot). Digits to the left of the point are increasing powers of ten, those to right are increasing negative powers of ten. Example 836.33, -45.009.

Fractions

A fraction is two quantities written one above the other, that shows how much of a a whole thing we have. For example we may have three quarters of a pizza:

$\frac{3}{4}$ of a pizza.

Normal Form (Scientific Notation)

For very large and very small numbers, decimal notation is not the most convenient. a number in normal form consists of two parts: a coefficient and an exponent (power of ten). For example, the distance to the sun is 93000000 miles. This can be more conveniently written as 93×106 miles. 93 is the coefficient and 6 is the exponent.

Sets

The collection of well-defined distinct objects is known as a set. The word well-defined refers to a specific property which makes it easy to identify whether the given object belongs to the set or not. The word 'distinct' means that the objects of a set must be all different.

For example

1. The collection of children in class VII whose weight exceeds 35 kg represents a set.
2. The collection of all the intelligent children in class VII does not represent a set because the word intelligent is vague. What may appear intelligent to one person may not appear the same to another person.

Elements of Set

The different objects that form a set are called the elements of a set. The elements of the set are written in any order and are not repeated. Elements are denoted by small letters.

Conventions for Sets

The following are the conventions that are used here:

- Sets are usually denoted by a capital letter.
- The elements of the group are usually represented by small letters (unless specified separately.)
- If 'a' is an element of 'A', or if a "belongs to" A, it is written in the conventional notion by the use of the Greek symbol ϵ (Epsilon) between them – $a \in A$.
- If b is not an element of Set A, b "does not belong to" A is written in the conventional notion by the use of the symbol ϵ (Epsilon with a line across it) between them – $a \in A$.
- Objects, elements, entities, members are all synonymous terms.

Representations of a Set

Representation of Sets and its elements is done in the following two ways.

Roster Form

In this form, all the elements are enclosed within braces {} and they are separated by commas (,). For example, a collection of all the numbers found on a dice $N = \{1, 2, 3, 4, 5, 6\}$.

Properties of roster form:

The order in which the elements are listed in the Roster form for any Set is immaterial. For example, V = {a, e, i, o, u} is same as V = {u, o, e, a, i}.

The dots at the end of the last element of any Set represent its infinite form and indefinite nature. For example, group of odd natural numbers = {1, 3, 5, ...}.

In this form of representation, the elements are generally not repeated. For example, the group of letters forming the word POOL = {P, O, L}.

More examples for Roster form of representation are:

- A = {3, 6, 9, 12}
- F = {2, 4, 8, 16, 32}
- H = {1, 4, 9, 16, ..., 100}
- L = {5, 25, 125, 625}
- Y = {1, 1, 2, 3, 5, 8, ...}.

Set Builder Form

In this form, all the elements possess a single common property which is NOT featured by any other element outside the Set. For example, a group of vowels in English alphabetical series.

The representation is done as follows. Let V be the collection of all English vowels, then – V = {x: x is a vowel in English alphabetical series.}

Properties of Roster form:

- Colon (:) is a mandatory symbol for this type of representation.
- After the colon sign, we write the common characteristic property possessed by ALL the elements belonging to that Set and enclose it within braces.
- If the Set doesn't follow a pattern, its Set builder form cannot be written.

More examples for Set builder form of representation for a Set:

$$D = \{x : \text{x is an integer and } -3 < x < 19\}$$

$$O = \{y : \text{y is a natural number greater than } 5\}$$

$$I = \{f : \text{f is a two – digit prime number less than } 1000\}$$

$$R = \{s : \text{s is a natural number such that sum of its digits is } 4\}$$

$$X = \{m : \text{m is a positive integer } < 40\}$$

Thus, these were some important points on Sets, what they are, how they are represented mathematically and the related properties.

Types of Sets

The different types of sets are as follows:

Empty Set

The set which is empty. This means that there are no elements in the set. This set is represented by ϕ or $\{\}$. An empty set is hence defined as,

If a set doesn't have any elements, it is known as an empty set or null set or void set. For e.g. consider the set,

$$P = \{x : \text{x is a leap year between 1904 and 1908}\}$$

Between 1904 and 1908, there is no leap year. So, P = ϕ. Similarly, the set

$$Q = \{y: y \text{ is a whole number which is not a natural number}, y \neq 0\}$$

0 is the only whole number that is not a natural number. If y ≠ 0, then there is no other value possible for y. Hence, Q = ϕ.

Singleton Set

If a set contains only one element, then it is called a singleton set.

$$A = \{x: x \text{ is an even prime number}\}$$

$$B = \{y: y \text{ is a whole number which is not a natural number}\}.$$

Finite Set

In this set, the number of elements is finite. All the empty sets also fall into the category of finite sets.

If a set contains no element or a definite number of elements, it is called finite set.

If the set is non-empty, it is called a non-empty finite set. Some examples of finite sets are:

$$A = \{x: x \text{ is a month in an year}\}; \text{ A will have 12 elements}$$

$$B = \{y: y \text{ is the zero of a polynomial } (x^4 - 6x^2 + x + 2)\}; \text{ B will have 4 zeroes.}$$

Infinite Set

Just contrary to the finite set, it will have infinite elements. If a given set is not finite, then it will be an infinite set.

For e.g.

$$A = \{x: x \text{ is a natural number}\}$$

There are infinite natural numbers. Hence, A is an infinite set.

B = $\{y: y \text{ is ordinate of a point on a given line}\}$; There are infinite points on a line. So, B is an infinite set.

Power Set

An understanding of what subsets are is required before going ahead with Power-set.

Definition: The power set of a set A is the set which consists of all the subsets of the set A. It is denoted by P(A).

For a set A which consists of n elements, the total number of subsets that can be formed is 2^n. From this, we can say that P(A) will have 2^n elements.

Sub Set

$$A = \{-9, 13, 6\}$$

Subsets of $A = \phi, \{-9\}, \{13\}, \{6\}, \{-9,13\}, \{13,6\}, \{6,-9\}, \{-9,13,6\}$

Definition: If a set A contains elements which are all the elements of set B as well, then A is known as the subset of B.

$$P(A) = \{\phi, \{-9\}, \{13\}, \{6\}, \{-9,13\}, \{13,6\}, \{6,-9\}, \{-9,13,6\}\}$$

Universal Set

This is the set which is the base for every other set formed. Depending upon the context, the universal set is decided. It may be a finite or infinite set. All the other sets are the subsets of the Universal set. It is represented by U.

For e.g. The set of real numbers is a universal set of integers. Similarly, the set of complex number is the universal set for real numbers.

Function

A Function from set A to set B is a Relation or a rule which associates or maps or images each and every element of set A with a element in set B.

A function is a special case or Relation in which each and every element of first set (A) is related with only one element of second set (B).

Denotation of Function

A function from set A to set B is denoted by,

$$f : A \rightarrow B$$

And read as "f is a function from set A to set B". We use f,g,h and F,G,H mainly to denote a function.

If we take any element of Set A and process it through the function (or use the rule of given function in it) then the element of set A is changed or imaged to another element which is an element of set B.

We denote the above statement by,

$$f(x) = y$$

Where "x" is a element of set "A" and "y" is the corresponding element of set "B".

This means "x" becomes "y" when it is processed through the function "f".

Likewise if we define a function "f" as "y is square of x":

Then,

$$f(x) = x^2$$

or, "x" becomes x^2 when it is processed through the function or rule "f".

So in the above example if set $A = \{1,2,3\}$ then when set "A" is processed through function "f" Set "A" is converted into another set(say B) where $B = \{2,4,9\}$.

A function can also be denoted by different graphical methods as given below.

If set $A = \{1,2,3\}$ is processed through a function "f" which is defined by "y is square of x" or,

$$f : A \rightarrow B \text{ or } y = f(x) = x^2$$

Then we can denote this function using following method:

a. Table Method

x	1	2	3
y	1	4	9

$f{:}A \rightarrow B$

b. Arrow Diagram method

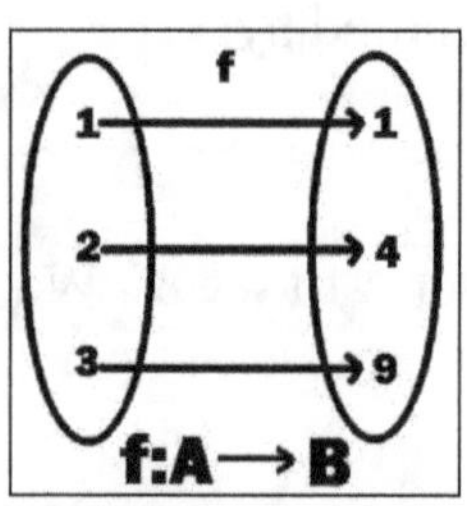

c. Graph method

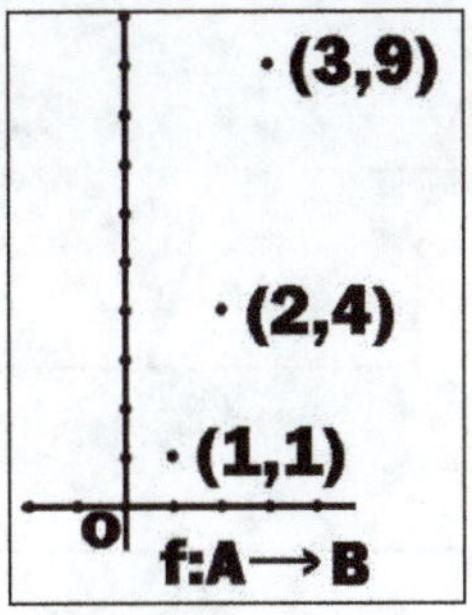

Domain and Range of a Function

Domain of a Function is the set of elements which are processed or to be processed by a function. And range is the set of elements which are produced after processing the domain of a Function.

For example:

If set $A=\{1,2,3\}$ and set A is to be processed by function "f" to produce another set $B=\{1,8,27\}$

or $f:A\rightarrow B$

Then set "A" is the domain of function "f" and set "B" is the range of function "f".

Types of Functions

One to One Function

A function $f:A\rightarrow B$ is One to One if for each element of A there is a distinct element of B. It is also known as Injective. Consider if $a_1\in A$ and $a_2\in B$, f is defined as $f:A\rightarrow B$ such that $f(a_1)=f(a_2)$

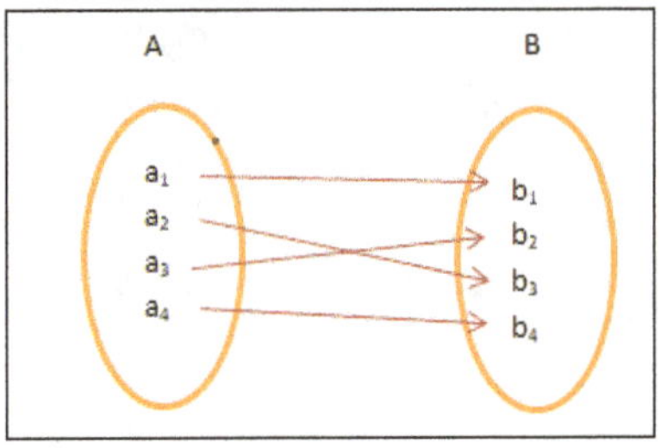

Many to One Function

It is a function which maps two or more elements of A to the same element of set B. Two or more elements of A have the same image in B.

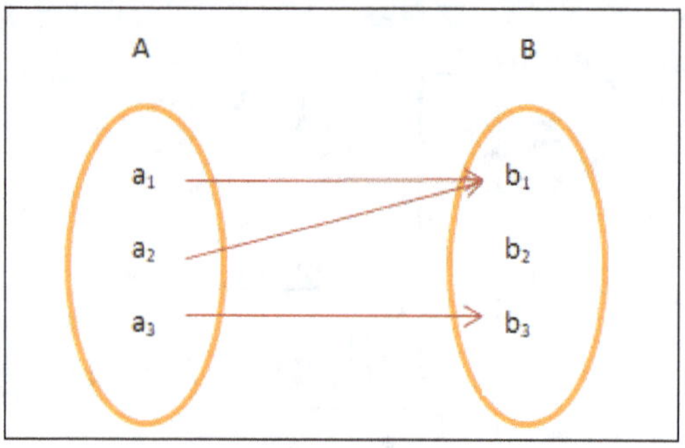

Onto Function

If there exists a function for which every element of set B there is (are) pre-image(s) in set A, it is Onto Function. Onto is also referred as Surjective Function.

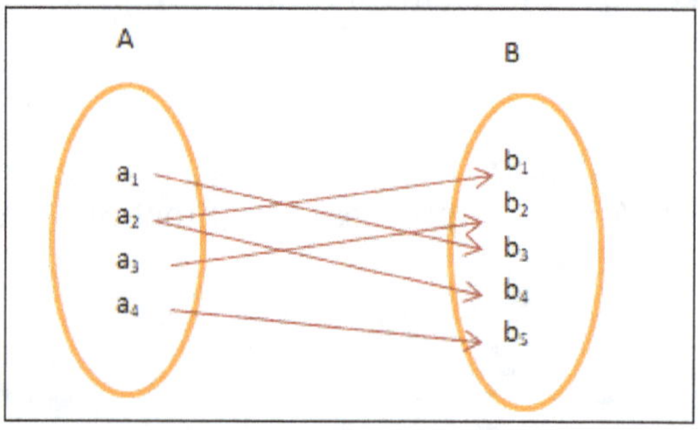

One – One and Onto Function

A function, f is One – One and Onto or Bijective if the function f is both One to One and Onto function. In other words, the function f associates each element of A with a distinct element of B and every element of B has a pre-image in A.

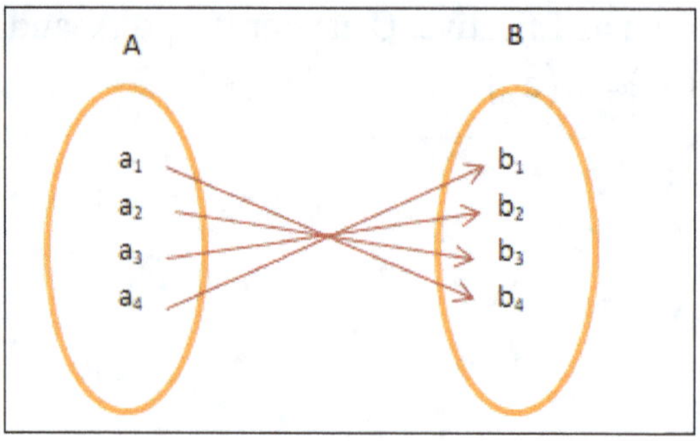

Other Types of Functions

A function is uniquely represented by its graph which is nothing but a set of all pairs of x and f(x) as coordinates. Let us get ready to know more about the types of functions and their graphs.

Identity Function

Let R be the set of real numbers. If the function $f: R \rightarrow R$ is defined as f(x) = y = x, for $x \in R$, then the function is known as Identity function. The domain and the range being R. The graph is always a straight line and passes through the origin.

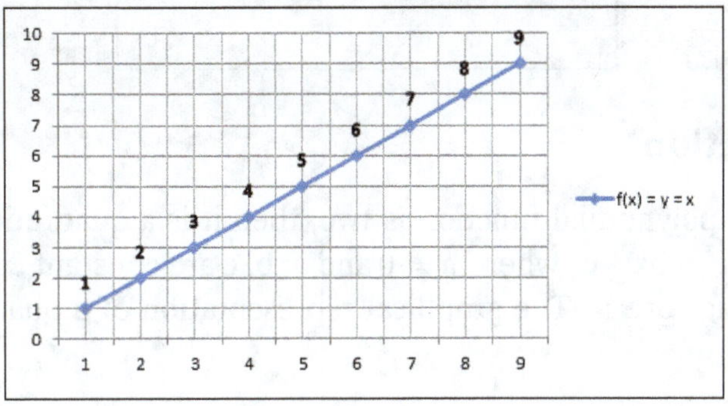

Constant Function

If the function $f: R \rightarrow R$ is defined as f(x) = y = c, for $x \in R$ and c is a constant in R, then such function is known as Constant function. The domain of the function f is R and its range is a constant, c. Plotting a graph, we find a straight line parallel to the x-axis.

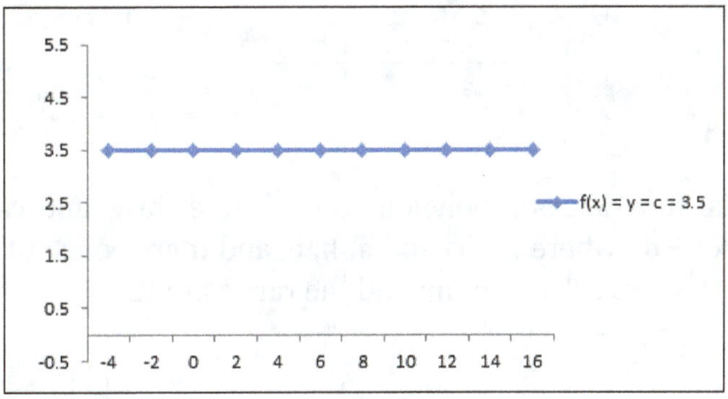

Polynomial Function

A polynomial function is defined by $y = a_0 + a_1x + a_2x^2 + \ldots + a_nx^n$, where n is a non-negative integer and $a_0, a_1, a_2, \ldots,_n \in R$. The highest power in the expression is the degree of the polynomial function. Polynomial functions are further classified based on their degrees:

- Constant Function: If the degree is zero, the polynomial function is a constant function.
- Linear Function: The polynomial function with degree one. Such as y = x + 1 or y = x or y = 2x – 5 etc. Taking into consideration, y = x – 6. The domain and the range are R. The graph is always a straight line.

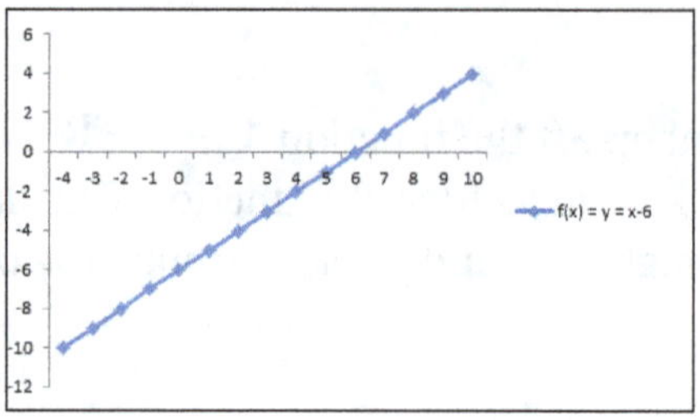

Quadratic Function

If the degree of the polynomial function is two, then it is a quadratic function. It is expressed as $f(x) = ax^2 + bx + c$, where $a \neq 0$ and a, b, c are constant & x is a variable. The domain and the range are R. The graphical representation of a quadratic function say, $f(x) = x^2 - 4$.

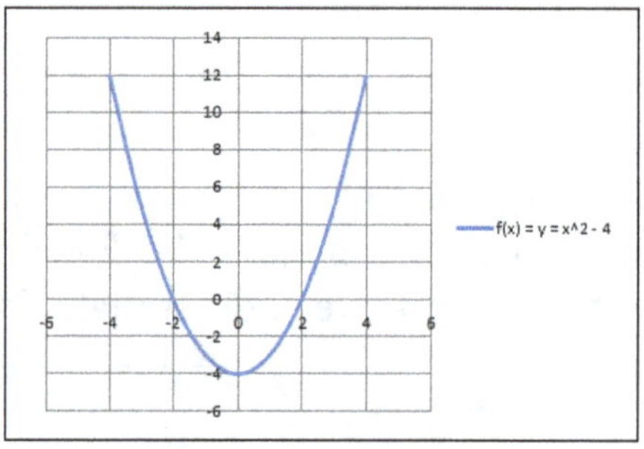

Cubic Function

A cubic polynomial function is a polynomial of degree three and can be denoted by $f(x) = ax^3 + bx^2 + cx + d$, where $a \neq 0$ and a, b, c, and d are constant & x is a variable. Graph for $f(x) = y = x^3 - 5$. The domain and the range are R.

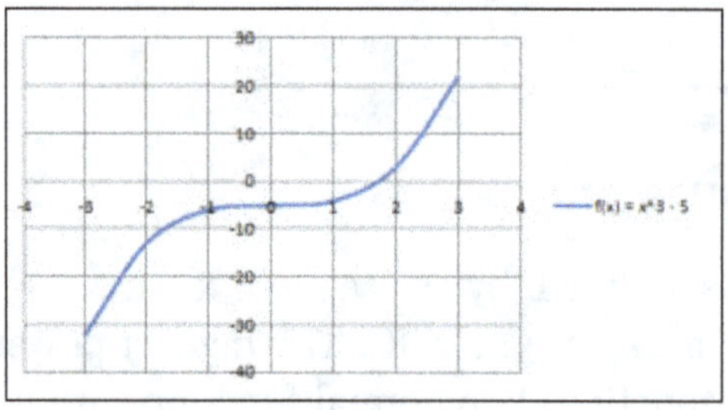

Rational Function

A rational function is any function which can be represented by a rational fraction say, f(x)/g(x) in which numerator, f(x) and denominator, g(x) are polynomial functions of x, where $g(x) \neq 0$. Let a function $f : R \rightarrow R$ is defined say, $f(x) = 1/(x + 2.5)$. The domain and the range are R. The Graphical representation shows asymptotes, the curves which seem to touch the axes-lines.

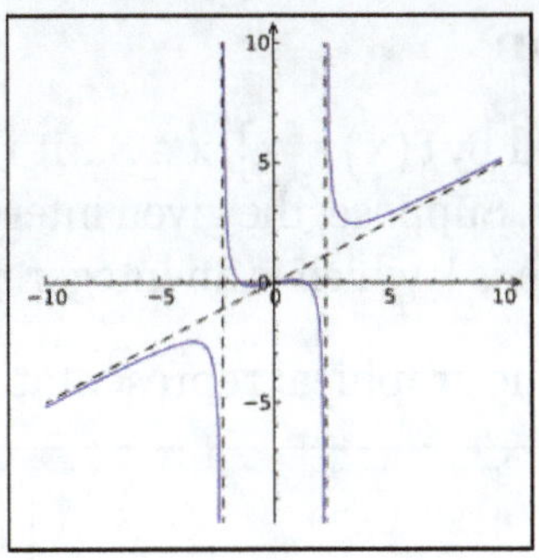

Modulus Function

The absolute value of any number, c is represented in the form of |c|. If any function $f: R \rightarrow R$ is defined by f(x) = |x|, it is known as Modulus Function. For each non-negative value of x, $f(x) = x$ and for each negative value of x, f(x) = -x, i.e.,

$$f(x) = \{x, \text{if } x \geq 0; -x, \text{if } x < 0.$$

Its graph is given as, where the domain and the range are R.

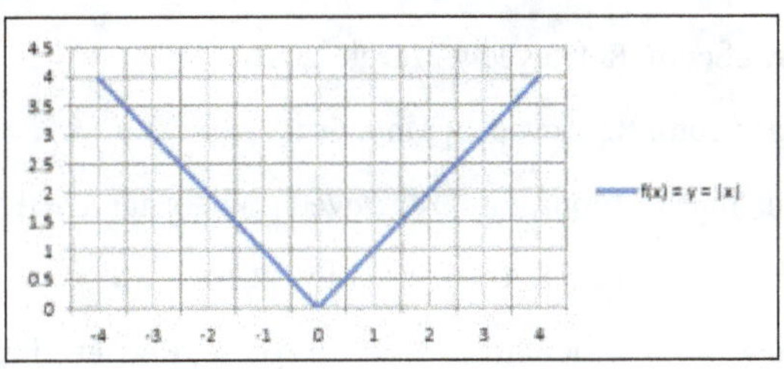

Signum Function

A function $f: R \rightarrow R$ defined by,

$$f(x) = \{1, \text{if } x > 0; 0, \text{if } x = 0; -1, \text{if } x < 0$$

Signum or the sign function extracts the sign of the real number and is also known as step function.

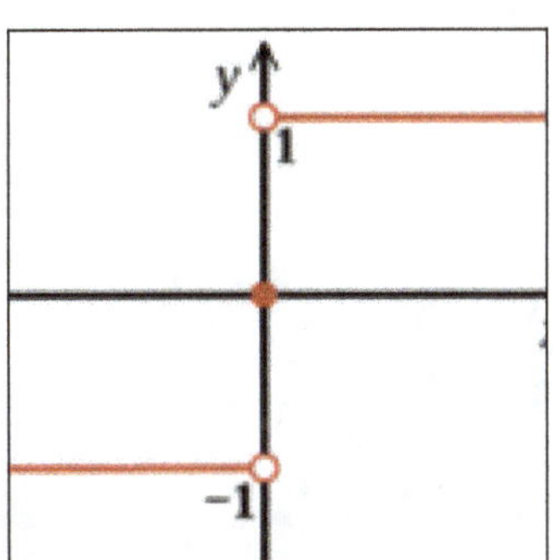

Greatest Integer Function

If a function $f : R \rightarrow R$ is defined by $f(x) = [x], x \in X$. It round-off to the real number to the integer less than the number. Suppose, the given interval is in the form of (k, k+1), the value of greatest integer function is k which is an integer.

For example: [-21] = 21, = 5. The graphical representation is given below.

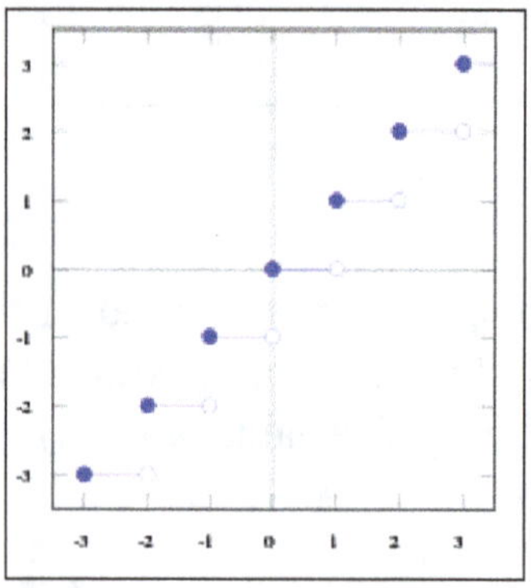

References

- Number: mathopenref.com, Retrieved 7 March, 2019
- Sets: math-only-math.com, Retrieved 12 May, 2019
- Introduction-to-sets, maths: toppr.com, Retrieved 28 February, 2019
- Types-of-sets, maths: byjus.com, Retrieved 3 July, 2019
- Introduction-to-functions, mathematics: sciencehq.com, Retrieved 22 April, 2019
- Types-of-functions, relations-and-functions, maths: toppr.com, Retrieved 12 January, 2019

2

Algebra

Algebra is a significant subset of mathematics that deals with the study of mathematical symbols and the rules for manipulating these symbols. It includes elementary equation solving as well as the study of abstractions such as fields, groups and rings. Elementary algebra and abstract algebra are the two main types of algebra. This chapter has been carefully written to provide an easy understanding of the varied facets of algebra.

Algebra is a branch of mathematics that substitutes letters for numbers. Algebra is about finding the unknown or putting real-life variables into equations and then solving them. Algebra can include real and complex numbers, matrices, and vectors. An algebraic equation represents a scale where what is done on one side of the scale is also done to the other and numbers act as constants.

Uses of Algebra

Algebra is widely used in many fields including medicine and accounting, but it can also be useful for everyday problem-solving. Along with developing critical thinking—such as logic, patterns, and deductive and inductive reasoning—understanding the core concepts of algebra can help people better handle complex problems involving numbers.

This can help them in the workplace where real-life scenarios of unknown variables related to expenses and profits require employees to use algebraic equations to determine the missing factors. For example, suppose an employee needed to determine how many boxes of detergent he started the day with if he sold 37 but still had 13 remaining. The algebraic equation for this problem would be:

$$x - 37 = 13$$

Where, the number of boxes of detergent he started with is represented by x, the unknown he is trying to solve. Algebra seeks to find the unknown and to find it here, the employee would manipulate the scale of the equation to isolate x on one side by adding 37 to both sides:

$$x - 37 + 37 = 13 + 37$$

$$x = 50$$

So, the employee started the day with 50 boxes of detergent if he had 13 remaining after selling 37 of them.

Types of Algebra

There are numerous branches of algebra, but these are generally considered the most important-

- Elementary: a branch of algebra that deals with the general properties of numbers and the relations between them.
- Abstract: deals with abstract algebraic structures rather than the usual number systems.
- Linear: focuses on linear equations such as linear functions and their representations through matrices and vector spaces.
- Boolean: used to analyze and simplify digital (logic) circuits. It uses only binary numbers, such as 0 and 1.
- Commutative: studies commutative rings—rings in which multiplication operations are commutative.
- Computer: studies and develops algorithms and software for manipulating mathematical expressions and objects.
- Homological: used to prove non-constructive existence theorems in algebra.
- Universal: studies common properties of all algebraic structures, including groups, rings, fields, and lattices.
- Relational: a procedural query language, which takes a relation as input and generates a relation as output.
- Algebraic number theory: a branch of number theory that uses the techniques of abstract algebra to study the integers, rational numbers, and their generalizations.
- Algebraic geometry: studies zeros of multivariate polynomials, algebraic expressions that include real numbers and variables.
- Algebraic combinatorics: studies finite or discrete structures, such as networks, polyhedra, codes, or algorithms.

Polynomial

In simple terms, polynomials are expressions comprising a sum of terms, where each term holding a variable or variables is elevated to a power and further multiplied by a coefficient. Amusingly, the simplest polynomials hold one variable.

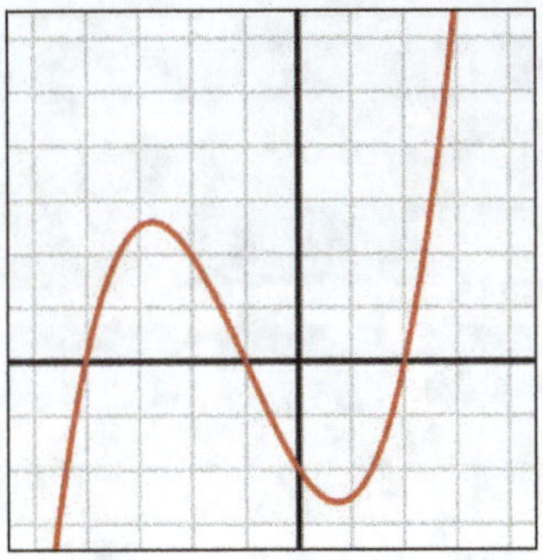

A single-variable polynomial having degree n has the following polynomial equation:

$$a_n x^n + a_{n-1} x^{n-1} + \ldots + a_2 x^2 + a_1 x^1 + a_0 x^0$$

In this, a's denote the coefficients whereas x denotes the variable. Since $x^1 = x$ and $x^0 = 1$ considering all complex numbers x. Therefore, the above expression can be shortened to:

$$a_n x^n + a_{n-1} x^{n-1} + \ldots + a_2 x^2 + a_1 x + a$$

When an nth-degree of single-variable polynomial equals to 0, then the resultant polynomial equation of degree 'n' acquires the following form:

$$a_n x^n + a_{n-1} x^{n-1} + \ldots + a_2 x^2 + a_1 x + a = 0.$$

Degree of a Polynomial

It is simply the greatest of the exponents or powers over the various terms present in the polynomial equation.

Example: Find the degree of a polynomial 7x – 5

In this, we observe that there are two terms in the mentioned polynomial. The first term is 7x, whereas the second term is -5. Hence, we define the exponent for each term. The exponent for the first term 7x = 1 and for the second term -5 = 0. Since the highest exponent is 1, therefore the degree of the polynomial 7x – 5 is also 1.

Types of Polynomials

1. Monomials – Monomials are the algebraic expression with one term, hence the name says "Mono"mial.
2. Binomials – Binomials are the algebraic expression with two unlike terms, hence the name "Bi"nomial.
3. Trinomials – Trinomials are the algebraic expression with three unlike terms, hence the name "Tri"nomial.

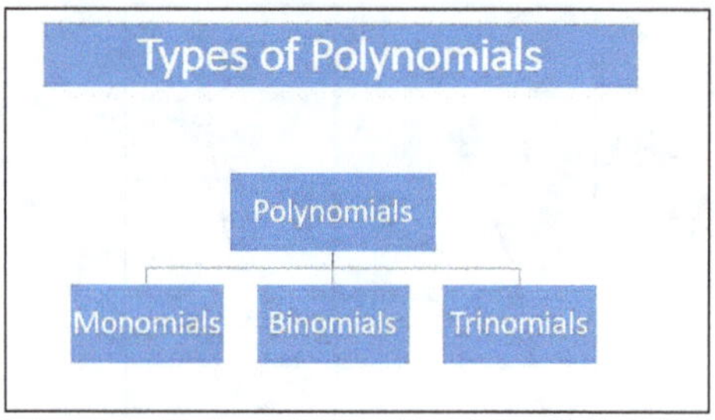

Types of Polynomial Equation

Let us try to get familiar with the different types of a polynomial equation which form the base to further learning.

- Zero Polynomial: Whenever in a given polynomial every coefficient value stays zero, then it is called as a zero polynomial. For example: 0 + 04 – 0.
- Monomial: It is an algebraic expression that contains only one term and is called as Monomial. In a simplistic form, it can be called as an expression that contains any count of like terms. For example: 2x, 4t, $21x^2y$, 9pq etc. Each of these expressions is monomial since they contain only one term.
- Binomial: It is an algebraic expression which comprises of two, unlike terms. For example, $3x + 4x^2$ is binomial since it contains two unlike terms, that is, 3x and $4x^2$. Also, $10pq + 13p^2q$ is also a Binomial. This is because it comprises two unlike terms, namely, 10pq and $13p^2q$.
- Trinomial: It is an algebraic expression that comprises three, unlike terms. For example- $3x+5x^2-6x^3$ is an active Trinomial. It is due to the presence of three, unlike terms, namely, $3x+5x^2$ and $6x^3$. Also, $12pq+4x^2-10$ is a trinomial, since it has three unlike terms- 12pq, $4x^2$ and 10.

Examples

Which of the following is a binomial?

a. $8*a+a$

b. $7a^2+8b+9c$

c. $3a*4b+2c$

d. $11a^2+11b^2$

solution: $11a^2+11b^2$

a. Will give 8a+a=9a which is monomial.

b. Is a trinomial.

c. Will give 24abc, which is a monomial.

d. Will give $11a^2+11b^2$, which is a binomial.

Zeroes of Polynomial

A polynomial having value zero (0) is known as zero polynomial. Actually, the term 0 is itself zero polynomial. It is a constant polynomial whose all the coefficients are equal to 0. For a polynomial, there may be few (one or more) values of the variable for which the polynomial may result in zero. These values are known as zeros of a polynomial. We can say that the whole.

If the coefficients of following the form of the polynomial:

$a_n x^n + a_{n-1} x^{n-1} + a_{n-2} x^{n-2} + \ldots + a_2 x^2 + a_1 x + a_0$ are zero, then it will become zero polynomial. i.e $a_n = a_{n-1} = a_{n-2} = \ldots = a_0 = 0$. Thus, the polynomial will become 0 and may be written as P(x)=0.

Zero Polynomial Function

The zero polynomial function is defined as the polynomial function with the value of zero. i.e. the function whose value is 0, is termed as a zero polynomial function. Zero polynomial does not have any nonzero term. It is represented as: P(x) = 0. Thus, we can say that a polynomial function which is equal to zero is called zero polynomial function. It also is known as zero map. The graph of the zero polynomial is X axis.

Zero Quadratic Polynomial

The quadratic polynomial having all the coefficients equal to zero is known as zero quadratic polynomial. The general term of a quadratic polynomial is: $P(x) = ax^2 + bx + c$. If in above quadratic polynomial, the coefficients are zero; i.e. a = b = c = 0, then the polynomial is termed as a zero quadratic polynomial.

Example: $0.x^2 + 0.x + 0$ is a zero quadractic polynomial whose values are zero.

Example: Find the additive identities of the following polynomials: 1) x-3 and 2) $x^2 - 3x + 5$.

Solution: 1) Additive identity = 0.x+0 and 2) Additive identity = $0.x^2 + 0.x + 0$

Finding Zeroes of a Polynomial

1. The zero of a polynomial is the value of the which polynomial gives zero. Thus, in order to find zeros of the polynomial, we simply equate polynomial to zero and find the possible values of variables.
2. Let P(x) be a given polynomial. To find zeros, set this polynomial equal to zero. i.e. P(x) = 0. Now, this becomes a polynomial equation. Solve this equation and find all the possible values of variables by factorizing the polynomial equation.

3. These are the values of x which make polynomial equal to zero; hence are called zeros of polynomial P(x). A number z is said to be a zero of a polynomial P(x) if and only if P(z) = 0.

Real and Complex Zeroes of Polynomials

When the roots of a polynomial are in the form of the real number, they are known as real zeros whereas complex numbers are written as a $\pm$ ib, where a is called real part and b is known as the imaginary part. The complex zeros are found in pairs, such as a + ib and a – ib.

Example: Find the zeroes of polynomial $6x^2 + 7x - 2$

Solution: To find zeros, set the polynomial equal to zero P(x)=0 i.e. $6x^2 + 7x - 2 = 0$

$6x^2 + 4x - 3x - 2 = 0$ then, $2\text{x}(3\text{x}+2) - 1(3\text{x}+2) = 0$

(3x+2)(2x-1)=0, $\text{x} = -\frac{2}{3}, \frac{1}{2}$.

Example: Find the zeroes of polynomial $(x-3)^2 + 4$

Solution: To find zeros, set the polynomial equal to zero P(x)=0 i.e. $(x-3)^2 + 4 = 0$

$(x-3)^2 = -4$ then, $\text{x} - 3 = \pm 2i$ and

Thus, two zeros are $3 + 2i$ and $3 - 2i$.

Algebraic Equation

In algebra, an equation can be defined as a mathematical statement consisting of an equal symbol between two algebraic expression that have the same value.

An equation of the type $f_n = 0$, where f_n is a polynomial of degree n in one or more variables ($n \geq 0$). An algebraic equation in one variable is an equation of the form

$$a_0x^n + a_1x^{n-1} + \ldots + a_n = 0 \cdot$$

Here n is a non-negative integer, $a_0, \ldots, a_n$ are given, the so-called coefficients of the equation, while x is an unknown which has to be found. It is assumed that the coefficients of the algebraic equation are not all equal to zero. If $a_0 \neq 0$, then n is called the degree of the equation.

The values of the unknown x which satisfy equation, i.e. the values which, if substituted for x, will convert this equation into an identity, are known as the roots of the equation ($a_0x^n + a_1x^{n-1} + \ldots + a_n = 0$), or as the roots of the polynomial,

$$f_n(x) = a_0x^n + a_1x^{n-1} + \ldots + a_n.$$

The roots of a polynomial are related to its coefficients by Viète's formulas. To solve an equation means to find all its roots contained in the range of values of the unknown(s) under consideration.

As far as applications are concerned, the most important case is that of coefficients and roots of an equation that are numbers of a certain kind (e.g. rational, real or complex). The case of the coefficients and roots being elements of an arbitrary field may also be considered.

If a given number (or element of a field) c is a root of the polynomial $f_n(x)$ then, in accordance with the Bezout theorem, $f_n(x)$ is divisible by $x - c$ without remainder. The division may be performed according to the Horner scheme.

A number (or element of a field) c is called a root of multiplicity k of a polynomial $f(x)$ (where k is a non-negative integer) if $f(x)$ is divisible by $(x-c)^k$, but is not divisible by $(x-c)^{k+1}$. Roots of multiplicity one are called simple roots of the polynomial, other roots are called multiple roots.

Each polynomial $f(x)$ of degree $n > 0$ with coefficients in a field P has at most n roots in this field P, each root being counted the number of times equal to its multiplicity (consequently, there are not more than n different roots).

In an algebraically closed field any polynomial of degree n has exactly n roots (counted according to their multiplicity). In particular, this statement also applies to the field of complex numbers.

Equation $a_0x^n + a_1x^{n-1} + \ldots + a_n = 0$ of degree n with coefficients from a field P is called irreducible over P if the polynomial is irreducible over this field, i.e. cannot be represented as the product of other polynomials of degrees lower than n over P. Otherwise, both the polynomial and the corresponding equation are called reducible. Polynomials of degree zero and zero itself are not considered to be reducible or irreducible. Whether a given polynomial is reducible or irreducible over a field P depends on the field in question. Thus, the polynomial $x^2 - 2$ is irreducible over the field of rational numbers, since it has no rational roots, but is reducible over the field of real numbers: $x^2 - 2 = (x+\sqrt{2})(x-\sqrt{2})$. Similarly, the polynomial $x^2 + 1$ is irreducible over the field of real numbers, but is reducible over the field of complex numbers. Only polynomials of the first degree are irreducible over the field of complex numbers, and any polynomial can be decomposed into linear factors. Only polynomials of the first degree and polynomials of the second degree without real roots are irreducible over the field of real numbers (and all polynomials can be decomposed into products of linear and irreducible quadratic polynomials). Irreducible polynomials of all degrees exist over the field of rational numbers; examples are the polynomials of the form $x^n + 2$. The irreducibility of a polynomial over the field of rational numbers can often be established by Eisenstein's criterion: If, for a polynomial of degree $n > 0$ with integral coefficients, there exists a prime number P such that the leading coefficient a_0 is not divisible by P, all the remaining coefficients are divisible by P, and the constant term a_n is not divisible by P^2, then this polynomial is irreducible over the field of rational numbers.

Let P be an arbitrary field. For each polynomial $f(x)$ of degree $n > 1$ that is irreducible over P, there exists an extension of P containing at least one root of $f(x)$; moreover, there exists a splitting field of $f(x)$, i.e. a minimal extension of P in which this polynomial can be decomposed into linear factors. Every field has an algebraically closed extension.

Solvability by Radicals

Any algebraic equation of degree not exceeding four is solvable by radicals.

Abel's theorem in modern formulation: Let ($a_0x^n + a_1x^{n-1} + \ldots + a_n = 0$) be an equation of degree $n > 4$ with coefficients $a_0, \ldots, a_n$; let K be an arbitrary field and let P be the field of rational functions in $a_0, \ldots, a_n$ with coefficients from K; then the roots of equation (lying in some extension of P) cannot be expressed in terms of the coefficients of this equation by a finite number of operations of addition, subtraction, multiplication and division (operations which are meaningful in P) and root extraction (which is meaningful in the extension of P). In other words, the general equation of degree $n > 4$ is unsolvable by radicals.

However, Abel's theorem does not contradict the fact that some algebraic equations with numerical coefficients (or with coefficients from a given field) are solvable by radicals. Some special equations of degree n are solvable by radicals (e.g. a two-term equation). A complete solution of the problem when an algebraic equation is solvable by radicals was given by E. Galois about the year 1830.

The fundamental theorem on the solvability of algebraic equations by radicals in Galois theory can be stated as follows. Let $f(x)$ be a polynomial with coefficients from a field K that is irreducible over K. Then, 1) if at least one root of the equation $f(x) = 0$ can be expressed by radicals in terms of the coefficients of this equation and if the exponents of the radicals are not divisible by the characteristic of K, then the Galois group of this equation is solvable over K; 2) conversely, if the Galois group of the equation $f(x) = 0$ is solvable over the field and the characteristic of K is zero or is higher than all orders of the constituent factors of this group, then all the roots of the equation can be represented by radicals in terms of its coefficients, all exponents of the roots

$a^{1/n}$ involved can be taken to be prime numbers, and the binomial equations $x^n - a = 0$ which correspond to these roots can be taken to be irreducible over the fields to which these radicals are adjoined.

This theorem was proved by Galois for the case in which K is the field of rational numbers; in this case all conditions involving the characteristic of the field K in the above formulation become superfluous.

Abel's theorem is a consequence of Galois' theorem, since the Galois group of equations of order n with coefficients (that are letters) over the field P of rational functions in the coefficients of the equation with coefficients from an arbitrary field K, is the symmetric group S_n, which is unsolvable for $n > 4$. For any $n > 4$ there exist equations of degree n with rational (and integral) coefficients that are unsolvable by radicals. An example of such an equation for $n = 5$ is the equation $x^5 - p^2 x - p = 0$, where P is a prime number. In Galois theory an algebraic equation is solved by reducing it to a chain of simpler equations, which are called resolvents of the original equation.

The solvability of equations by radicals is closely connected with problems involving geometrical constructions with ruler and compasses; in particular, with the problem of the division of the circle into n equal parts.

Algebraic Equations in One Unknown with Numerical Coefficients

Methods of approximate calculations (e.g. the parabola method) are generally used to find the roots of algebraic equations of degree higher than two with coefficients from the field of real or complex numbers. It is convenient to begin by getting rid of the multiple roots. A number c is a root of multiplicity k of a polynomial $f(x)$ if and only if the polynomial and its derivatives up to the order $k-1$, inclusive, vanish if $x = c$, and if $f^{(k)}(c) \neq 0$. If $f(x)$ is divided by the greatest common divisor $d(x)$ of this polynomial and its derivative, then the resulting polynomial has the same roots as $f(x)$, but only of multiplicity one. It is further possible to construct the polynomials whose simple roots are all the roots of the polynomial $f(x)$ of equal (given) multiplicity. A polynomial has multiple roots if and only if its discriminant vanishes.

The determination of the number of roots and bounds on their size are frequently occurring problems.

$$1 + \frac{\max_{i>0} |a_i|}{|a_0|}$$

The number, can be taken as an upper bound for the modulus of each root (both real and complex) of the algebraic equation with arbitrary complex coefficients. The Newton method usually yields a more exact bound if the coefficients are real. The determination of a lower bound for the positive roots, and of upper and lower bounds for the negative roots, are reduced to the determination of an upper bound for the positive roots.

The simplest method to determine the number of real roots is to use the Descartes theorem. If it is known that all the roots of a given polynomial are real (for example, for the characteristic polynomial of a real symmetric matrix), then Descartes' theorem yields the exact number of roots. By considering the polynomial $f(-x)$, the number of negative roots of $f(x)$ can be found using the same theorem. The exact number of real roots located in a given interval (in particular, the number of all real roots) of a polynomial with real coefficients without multiple roots can be found by the Sturm theorem. Descartes' theorem is a special case of the Budan–Fourier theorem, which yields an estimate from above for the number of real roots of a polynomial with real coefficients lying in a certain fixed interval.

It is sometimes desirable to find roots of a special type. For instance, Hurwitz' criterion is a necessary and sufficient condition for all the roots of an equation (with complex coefficients) to have negative real parts.

There exists a method for calculating all rational roots of a polynomial with rational coefficients. A polynomial $f(x)$ with rational coefficients has the same roots as the polynomial $g(x)$ with integral coefficients obtained from $f(x)$ by multiplication by a common multiple of all denominators of the coefficients of $f(x)$. The only rational roots of a polynomial $g(x) = b_0 x^n + b_1 x^{n-1} + \ldots + b_n$, $b_n \neq 0$, with integral coefficients are found among the irreducible fractions p / q in which P is a divisor of the number b_n and q is a divisor of the number b_0 (and only those fractions among them such that, for any integer m, the number $g(m)$ is divisible by $p - mq$). If $b_0 = 1$, then all rational roots of $g(x)$ (if any) are integers, divisors of the constant term, and can be found by trial and error.

Systems of Algebraic Equations

Concerning systems of algebraic equations of the first degree.

A system of any two algebraic equations of any degree in two unknowns x and y may be written as:

$$\left.\begin{aligned} f(x, y) &= a_0(x)y^n + a_1(x)y^{n-1} + \ldots + a_n(x) = 0, \\ g(x, y) &= b_0(x)y^s + b_1(x)y^{s-1} + \ldots + b_s(x) = 0, \end{aligned}\right\}$$

Where $a_i(x), b_j(x)$ are polynomials in one unknown x. If a certain numerical value is assigned to x, a system of two equations in one unknown y with constant coefficients a_i, b_j is obtained. The resultant of this system is the following determinant:

$$R(f,g)=\begin{vmatrix} a_0 & a_1 & \cdots & a_n & 0 & 0 & \cdots & 0 \\ 0 & a_0 & \cdots & \cdots & a_n & \cdots & \cdots & \cdots \\ \cdots & \cdots & \cdots & \cdots & \cdots & 0 & \cdots & 0 \\ 0 & 0 & \cdots & \cdots & 0 & a_0 & \cdots & a_n \\ b_0 & b_1 & \cdots & b_s & 0 & 0 & \cdots & 0 \\ 0 & b_0 & \cdots & \cdots & b_s & \cdots & \cdots & \cdots \\ \cdots & \cdots & \cdots & \cdots & \cdots & 0 & \cdots & 0 \\ 0 & 0 & \cdots & \cdots & 0 & b_0 & \cdots & b_s \end{vmatrix}$$

(There are s rows of a's and n rows of b's in this determinant.) The following statement is true: A number x_0 is a root of the resultant $R(f, g)$ if and only if the polynomials $f(x_0, y)$ and $g(x_0, y)$ have a common root y_0 or if the two leading coefficients $a_0(x_0)$ and $b_0(x_0)$ are equal to zero.

Thus, in order to solve the system one must find all roots of the resultant $R(f, g)$, substitute each one of these roots into the system and find the common roots of these two equations in one unknown y. One must also find the common roots of the two polynomials $a_0(x)$ and $b_0(x)$, substitute them into, and verify if the resulting equations in one unknown y have common roots. In other words, the solution of a system of two algebraic equations in two unknowns is reduced to the solution of one equation in one unknown and the calculation of the common roots of two equations in one unknown (the common roots of two or more polynomials in one unknown are the roots of their largest common divisor).

Types of Algebraic Equations

Linear Equation

$$\begin{cases} 3x + 2y = 6 \\ 2x + 3y = 5 \end{cases}$$

$$\begin{cases} 2x + y = 3 \\ x + 2y = -1 \end{cases}$$

Linear equations are those where each term is either a constant or the product of a single variable and a constant. If there are two variables, the graph of the linear equation is always going to be a straight line. As a general rule, a linear equation looks like this:

$$y = mx + c, m \neq\neq 0$$

In this example, m is known as slope and c represents that point on which it cut the y axis.

In linear equations with different variables,

The equation with only one variable: an equation which has only one variable. Examples include the following,

- $8a - 8 = 0$
- $9a = 72$

The equation which has two variables: an equation that has only two types of variables. Examples include the following,

- $9a + 6b - 82 = 0$
- $7x + 7y = 12$
- $8a - 8d = 74$

The equation that has three variables: this is an equation with only three types of variables in the equation. Examples include the following,

- $13a - 8b + 31c = 74$
- $5x + 7y - 6z = 12$
- $6p + 14q - 74 + 82 = 0$·

Quadratic Equation

A quadratic equation is written in the **Standard Form**,

$$ax^2 + bx + c = 0$$

where a, b, and c are real numbers and $a \neq 0$.

Examples:

$x^2 - 7x + 12 = 0$ (standard form)

$x(x+7) = 0$

$3x^2 + 4x = 15$

A quadratic equation is a second-degree equation whereby one variable contains the variable that has an exponent of two. An example and the general form is shown below,

$$ax^2 + bx + c = 0, a \neq\neq 0$$

Other examples include,

- $5a^2 - 5a = 35$
- $8x^2 + 7x - 75 = 0$
- $4y^2 + 14y - 8 = 0$·

Cubic Equation

A cubic equation is a polynomial equation whereby the highest sum of exponents of the variables in any term is equal to three. In other words, it is an equation involving a cubic polynomial; i.e., one of the form. It has the following form:

$$ax^3 + bx^2 + cx + d = 0 \text{ where} \neq 0.$$

Exponential Equation

Exponential equations have variables in the place of exponents, and can be solved using this property: axax = ayay => x = y.

Examples include the following:

- 4x = 0
- 8x = 32
- ab = 0 (where "a" is base and "b" is the exponent).

Quartic Equation

Quartic equations are equations of the fourth degree and an equation that equates a quartic polynomial to zero, using this form:

$$f(x) = ax^4 + bx^3 + cx^2 + dx + e = 0 \text{ where } a \neq 0$$

The derivative of a quartic function is a cubic function.

Quintic Equation

A quintic equation is a polynomial equation in which five is the highest power of the variable. The formula used is,

$$ax^5 + bx^4 + cx^3 + dx^2 + ex + f = 0$$

Examples include the following:

x^5+x^3+x

$y^5+y^4+y^3+y^2+y+1$.

Radical Equation

Radical equations are those that have a maximum exponent on the variable that is 12 and which have more than one term. It can also be said that a radical equation is one whereby the variable is lying inside a radical symbol, usually in the form of a square root.

Examples include the following,

- + 10 = 26
- + x − 1.

Fundamental Theorem of Algebra

The Fundamental theorem of algebra states that any non-constant polynomial with complex coefficients has at least one complex root. The theorem implies that any polynomial with complex coefficients of degree n has n complex roots, counted with multiplicity. A field F with the property that every non-constant polynomial with coefficients in F has a root in F is called algebraically closed, so the fundamental theorem of algebra states that,

The field $\mathbb{C}$ of complex numbers is algebraically closed.

Example

The polynomial x^2+1 has no real roots, but it has two complex roots i and $-i$.

The polynomial x^2+1 has two complex roots, namely $\pm\frac{1-i}{\sqrt{2}}$.

One might expect that polynomials with complex coefficients have issues with nonexistence of roots similar to those of real polynomials; that is,

$$x^3+ix^2-(1+\pi i)x-e$$

it is not unreasonable to guess that some polynomial like will not have a complex root, and finding such a root will require looking in some larger field containing the complex numbers. The fundamental theorem of algebra says that this is not the case: all the roots of a polynomial with complex coefficients can be found living inside the complex numbers already.

Factoring

This topic gives a more precise statement of the different equivalent forms of the fundamental theorem of algebra. This requires a definition of the multiplicity of a root of a polynomial.

The multiplicity of a root r of a polynomial $f(x)$ is the largest positive integer k such that $(x-r)^k$ divides $f(x)$. Equivalently, it is the smallest positive integer k such that $f^{(k)}(r) \neq 0$, where $f^{(k)}$ denotes the kth derivative of f.

Theorem

Let F be a field. The following are equivalent:

1. Every non-constant polynomial with coefficients in F has a root in F.
2. Every non-constant polynomial of degree n with coefficients in F has nn roots in F, counted with multiplicity.
3. Every non-constant polynomial with coefficients in F splits completely as a product of linear factors with coefficients in F.

Proof

Clearly F $(3) \Rightarrow (2) \Rightarrow (1)$, so the only nontrivial part is $(1) \Rightarrow (3)$. To see this, induct on the degree n of $f(x)$. The base case $n=1$ is clear. Now suppose the result holds for polynomials of degree $n-1$. Then let $f(x)$ be a polynomial of degree n. By (1), $f(x)$ has a root a.a. A standard division algorithm argument shows that $x-a$ is a factor of $f(x)$:

Divide $f(x)$ by $x-a$ to get $f(x) = (x-a)q(x)+r$, where rr is a constant polynomial. Plugging in aa to both sides gives $0=(a-a)q(a)+r$, so $r=0$. So $f(x)=(x-a)q(x)$. But $q(x)$ is a polynomial of degree $n-1$, so it splits into a product of linear factors by the inductive hypothesis. Therefore $f(x)$ does as well. So the result is proved by induction.

The fundamental theorem of algebra says that the field $\mathbb{C}$ of complex numbers has property (1), so by the theorem above it must have properties (1), (2), and (3).

Example:

If $f(x)=x^4-x^3-x+1$, then complex roots can be factored out one by one until the polynomial is factored completely: f $f(1)=0$, so $x^4-x^3-x+1=(x-1)(x^3-1)$.

Then 1 is a root of x^3-1, so

$$x^4-x^3-x+1=(x-1)(x-1)(x^2+x+1).$$

And now x^2+x+1 has two complex roots, namely the primitive third roots of unity ω and ω^2, where $\omega=e^{2\pi i/3}$.

$$x^4-x^3-x+1=(x-1)^2(x-\omega)(x-\omega^2).$$

So here are three distinct roots, but four roots with multiplicity, since the root 1 has multiplicity 2.

Applications of the Theorem

The ability to factor any polynomial over the complex numbers reduces many difficult nonlinear problems over other fields (e.g. the real numbers) to linear ones over the complex numbers. For example, every square matrix over the complex numbers has a complex eigenvalue, because the characteristic polynomial always has a root. This is not true over the real numbers, e.g. the matrix:

$$\begin{pmatrix} 0 & 1 \\ -1 & 0 \end{pmatrix}$$

Which rotates the real coordinate plane by 90°, has no real eigenvalues.

Another general application is to the field of algebraic geometry, or the study of solutions to polynomial equations. The assumption that the coefficients of the polynomial equations lie in an algebraically closed field is essential for simplifying and strengthening the theory, as it guarantees that the field is "big enough" to contain roots of polynomials. For example, the set of complex solutions to a polynomial equation with real coefficients often has more natural and useful properties than the set of real solutions.

Another application worth mentioning briefly is to integration with partial fractions. Over the real numbers, there are awkward cases involving irreducible quadratic factors of the denominator. The algebra is simplified by using partial fractions over the complex numbers (with the caveat that some complex analysis is required to interpret the resulting integrals).

Polynomials over the Real Numbers

Let $p(x)$ be a polynomial with real coefficients. It is true that $p(x)$ can be factored into linear factors over the complex numbers, but the factorization is slightly more complicated if the factors are required to have real coefficients.

For instance, the polynomial x^2+1 can be factored as $(x-i)(x+i)$ over the complex numbers, but over the real numbers it is irreducible: it cannot be written as a product of two nonconstant polynomials with real coefficients.

Theorem

Every polynomial $p(x)$ with real coefficients can be factored into a product of linear and irreducible quadratic factors with real coefficients.

Proof

Induct on n. The base cases are $n=1$ and $n=2$, which are trivial. Now suppose the theorem is true for polynomials of degree $n-2$ and $n-1$. Let f(x) be a polynomial of degree n, and let f(x) be a complex root of $f(x)$ (which exists by the fundamental theorem of algebra). There are two cases:

- If a is real, then $f(x)=(x-a)q(x)$ for a polynomial $q(x)$ with real coefficients of degree $n-1$.

By the inductive hypothesis, $q(x)$ can be factored into a product of linear and irreducible quadratic factors, so $f(x)$ can as well.

- If f(x) a is not real, then let $\overline{a}$ be the complex conjugate of a. Note that $\overline{a} \neq a$. Write $f(x)=c_n x^n+\cdots+c_1 x+c_0$, then,

$$\begin{aligned}
\overline{f(x)} &= \overline{c_n x^n+\cdots+c_1 x+c_0} \\
&= \overline{c_n x^n}+\cdots+\overline{c_1 x}+\overline{c_0} \\
&= \overline{c_n}\,\overline{x}^n+\cdots+\overline{c_1}\,\overline{x}+\overline{c_0} \\
&= c_n \overline{x}^n+\cdots+c_1 \overline{x}+c_0 \\
&= f(\overline{x})
\end{aligned}$$

By properties of the complex conjugate (and because the c_i are real numbers). So if $f(a)=0$, then $f(\overline{a})=\overline{f(a)}=\overline{0}=0$. The conclusion is that non-real roots of polynomials with real coefficients come in complex conjugate pairs.

Write $f(x)=(x-a)q(x)$, where $q(x)$ has complex coefficients, and plug in $\overline{a}$ to both sides. Then $q(\overline{a})=0$. (This is where the argument uses that $\overline{a} \neq a$ So, $q(x)=(x-\overline{a})h(x)$, $f(x)=(x-a)(x-\overline{a})h(x)$ so. Write the product of the first two factors as $q(x)$, then $q(x)$ is a quadratic irreducible polynomial with real coefficients. Since $q(x)$ divides $f(x)$ over the complex numbers, and both polynomials are real, $q(x)$ must divide $f(x)$ over the real numbers. (Proof: use the division algorithm over the real numbers, $f=gj+k$, with $k=0$ or $deg(k)<deg(g)$, and then over the complex numbers g divides ff and gj, so must divide k; so $k=0$.)

So $h(x)$ is a polynomial of degree $n-2$ with real coefficients, which factors as expected by the inductive hypothesis, so f(x) $f(x)$ does as well. This completes the proof.

Proof of the Theorem

There are no "elementary" proofs of the theorem. The easiest proofs use basic facts from complex analysis. Here is a proof using Liouville's theorem that a bounded holomorphic function on the entire plane must be constant, along with a basic fact from topology about compact sets.

Let $p(z) = a_n z^n + \cdots + a_0$ be a polynomial with complex coefficients, and suppose that $p(z) \neq 0$ everywhere. (So of course $a_0 \neq 0$.) Then $\frac{1}{p(z)}$ is holomorphic everywhere. Now $\lim_{z\to\infty} p(z) = \infty$, for instance, because

$$|p(z)| \geq |a_n||z|^n - \left(|a_{n-1}||z|^{n-1} + \cdots + |a_0|\right)$$

By the triangle inequality. So for large enough $|z|$ say $|z| > R, p(z)| > |a_0|$.

But in the disc $|z| \leq R$, the function $|p(z)|$ attains its minimum value (because the disc is compact). Call this value m.

$$m > 0.$$

Then $|p(z)| > \min(m, |a_0|)$ for all Z, so.

$$\left|\frac{1}{p(z)}\right| < \frac{1}{\min(m, |a_0|)}$$

For all z, so it is a bounded holomorphic function on the entire plane, so it must be constant by Liouville's theorem. But then $p(z)$ is constant.

References

- Definition-of-algebra: thoughtco.com, Retrieved 17 June, 2019
- Polynomial-and-its-types, algebraic-expressions-and-identities, maths: toppr.com, Retrieved 13 August, 2019
- Zeroes-of-polynomial: mathstips.com, Retrieved 23 February, 2019
- Algebraic_equation: encyclopediaofmath.org, Retrieved 20 May, 2019
- Types-of-equations: popoptiq.com, Retrieved 10 January, 2019
- Fundamental-theorem-of-algebra: brilliant.org, Retrieved 2 March, 2019

3

Geometry

The branch of mathematics that is concerned with the questions of size, relative position of figures, shape and the properties of space is known as geometry. Euclidean geometry, Differential geometry, Algebraic geometry and Solid geometry are the various types of geometry. The topics elaborated in this chapter will help in gaining a better perspective about the various types of geometry.

Geometry is a branch of mathematics that studies the size, shape, and position of 2-dimensional shapes and 3-dimensional figures.

The main concepts in geometry are lines and segments, shapes and solids (including polygons), triangles and angles, and the circumference of a circle. In Euclidean geometry, angles are used to study polygons and triangles.

As a simple description, the fundamental structure in geometry–a line–was introduced by ancient mathematicians to represent straight objects with negligible width and depth. Plane geometry studies flat shapes like lines, circles, and triangles, pretty much any shape that can be drawn on a piece of paper. Meanwhile, solid geometry studies three-dimensional objects like cubes, prisms, cylinders, and spheres.

Fundamental Geometric Figures

When studying geometry, we will deal with several fundamental figures, including points, lines, and planes. A point can be thought of as an infinitesimally small sphere; that is, if you consider a "round" object such as a tennis ball, imagine that ball getting smaller and smaller until it is so small that it cannot be measured or even seen--this is akin to a point. A point has no width, length, or depth. A point can also be thought of as simply a location; a location does not have any physical dimension such as length or width, and yet we know intuitively how to speak of locations: "no trespassing beyond this point," for instance. Even though a point has no measurable size, we still need a way to represent it when studying geometry. Thus, we typically represent points as small dots, such as those shown below.

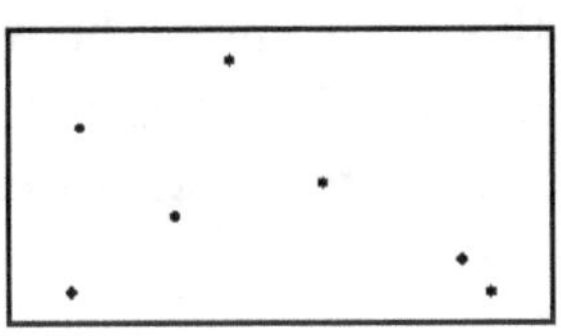

A line has slightly more character than a point: it is a figure that extends infinitely in one dimension. (A dimension can simply be thought of as a pair of opposite directions: forward and backward, left and right, or up and down, for example.) A line has infinite length, but it has no thickness. Imagine a piece of string that is pulled tight from two points that are immeasurably far away, and imagine that the thickness of the string decreases until the string is invisible: this string would then be a line. Because we can't show infinite distances on a piece of paper (or anywhere else, for that matter), we use arrows when drawing lines to indicate that the figure extends indefinitely. Below is an example of a line.

Naturally, we can conceive of a line that extends to infinity (that is, extends indefinitely) in only one direction but has an end point in the other direction; this figure is called a ray. Imagine the sun shining in the sky--a beam of light that originates from the sun can extend indefinitely away from it. The word "ray" in this case is closely linked with the geometric figure called a "ray." A ray is illustrated below.

If the line instead has two endpoints, we call it a line segment, which is illustrated below.

In some cases, it is helpful to show the endpoints of the ray or line segment more clearly using dots. The figures are the same, however.

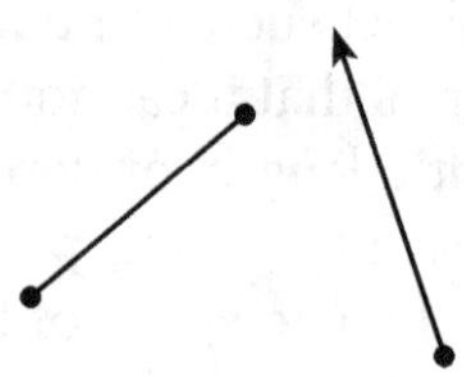

A plane is a geometric figure that extends to infinity in two dimensions. As such, we must show it in three dimensions, but you can imagine a plane as an even surface whose edges are so far away that they could never be reached. Planes are often drawn as parallelograms (four-sided figures similar to rectangles), since their infinite extent cannot be represented. We will draw the plane with a dashed border to differentiate it from a parallelogram.

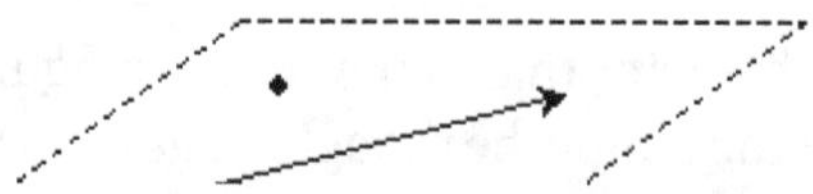

Because we will mostly deal with (two-dimensional) planar geometry, we will not deal much with planes. However, that we are in a sense dealing with a plane in planar geometry: we do all our drawing, moving, and analyzing of figures located entirely on that one plane. Consider, for instance, the collection of basic figures shown on the plane below.

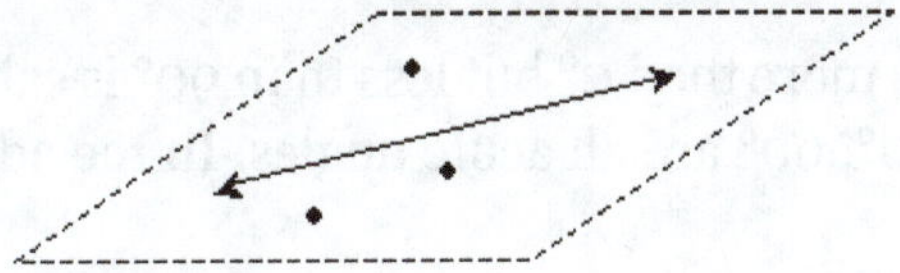

Plane Geometry

Plane geometry is the branch of geometry that deals with geometric figures, that is, collections of points that all lie in the same plane (coplanar). Although the words "point" and "plane" are undefined concepts, for elementary applications the intuitive meanings will serve: a point is a location, and a plane is a flat surface.

Angle

In geometry, an angle can be defined as the figure formed by two rays meeting at a common end point.

An angle is represented by the symbol ∠. Here, the angle below is ∠AOB.

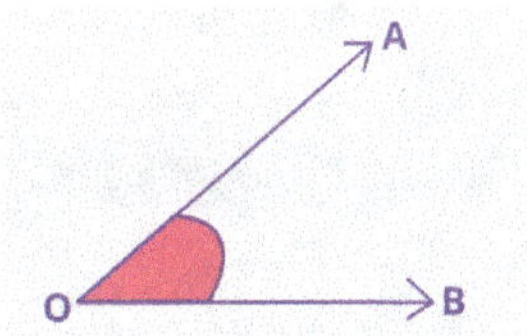

Angles are measured in degrees, using a protractor.

Parts of an Angle

- Arms: The two rays joining to form an angle are called arms of an angle. Here, OA and OB are the arms of the ∠AOB.
- Vertex: The common end point at which the two rays meet to form an angle is called the vertex. Here, the point O is the vertex of ∠AOB.

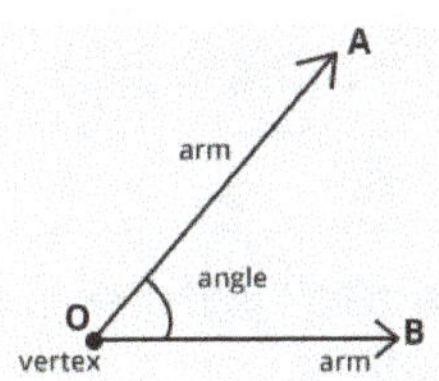

Classification of Angles

Classification of angles on the basis of their degree measures are given below:

Acute Angle

An angle whose measure is more than 0° but less than 90° is called an acute angle. Angles having magnitudes 30°, 40°, 60° are all acute angles. In the adjoining figure, ∠XoY represents an acute angle.

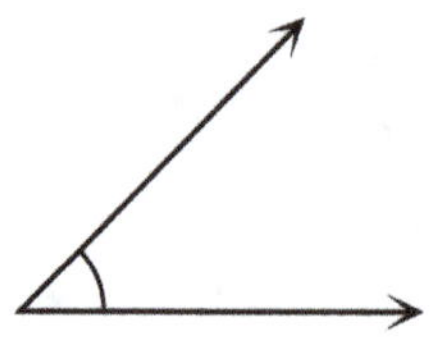

∠XoY < 90°

Right Angle

An angle whose measure is equal to 90° is called a right angle. In the adjoining figure ∠ABC represents a right angle.

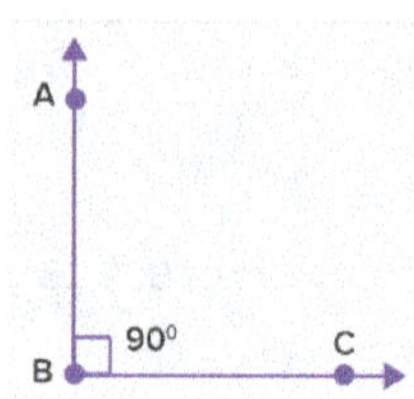

∠ABC = 90°

Obtuse Angle

An angle whose measure is more than 90° but less than 180° is called an obtuse angle. In the adjoining figure, ∠XYZ represents an obtuse angle.

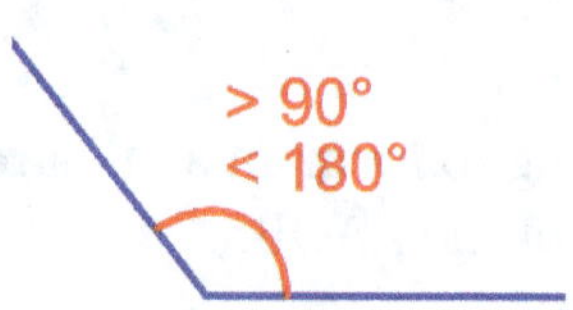

∠XYZ > 90°

∠XYZ < 180°

Straight Angle

An angle whose measure is equal to 180° is called a straight angle. In the adjoining figure, ∠XOY represents a straight angle.

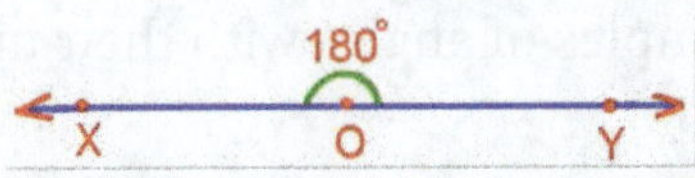

∠XOY = 180°

Reflex Angle

An angle whose measure is more than 180° but less than 360° is called a reflex angle. In the adjoining figure, ∠POQ is a reflex angle. Angles having magnitudes 220°, 250°, 310° are all reflex angles.

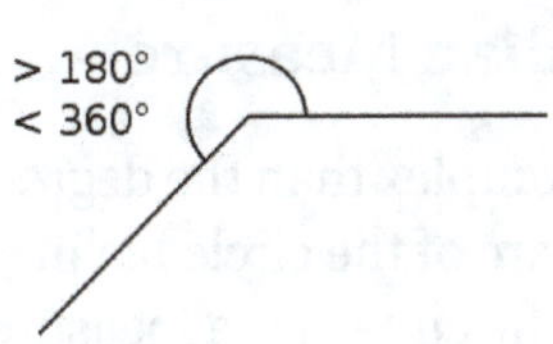

∠POQ > 180°

∠POQ < 360°

Complete Angle

An angle whose measure is equal to 360° is called a complete angle. In the adjoining figure, ∠BOA represents a complete angle.

60 minutes = 1 revolution = 1 complete angle.

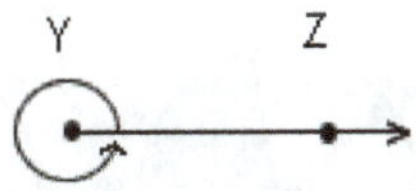

These are the adjoining figures of the classification of angles on the basis of their degree measures.

Angle Measurement – Degree Measure

A complete revolution, i.e. when the initial and terminal sides are in the same position after rotating clockwise or anticlockwise, is divided into 360 units called *degrees*. So, if the rotation from the initial side to the terminal side is $\left(\frac{1}{360}\right)$th of a revolution, then the angle is said to have a measure of one degree. It is denoted as 1°.

We measure time in hours, minutes, and seconds, where 1 hour = 60 minutes and 1 minute = 60 seconds. Similarly, while measuring angles,

- 1 degree = 60 minutes denoted as $1^\circ = 60'$
- 1 minute = 60 seconds denoted as $1' = 60''$

Here are some additional examples of angles with their measurements:

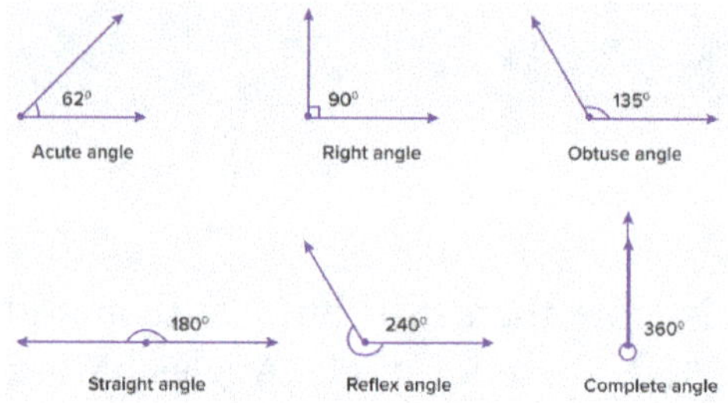

Angle Measurement – Radian Measure

Radian measure is slightly more complex than the degree measure. Imagine a circle with a radius of 1 unit. Next, imagine an arc of the circle having a length of 1 unit. The angle subtended by this arc at the centre of the circle has a measure of 1 radian. Here is how it looks:

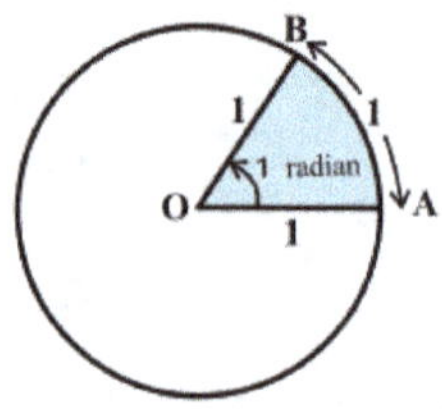

Here are some more examples of angles that measure -1 radian, $1\frac{1}{2}$ radian, and $-1\frac{1}{2}$ radian.

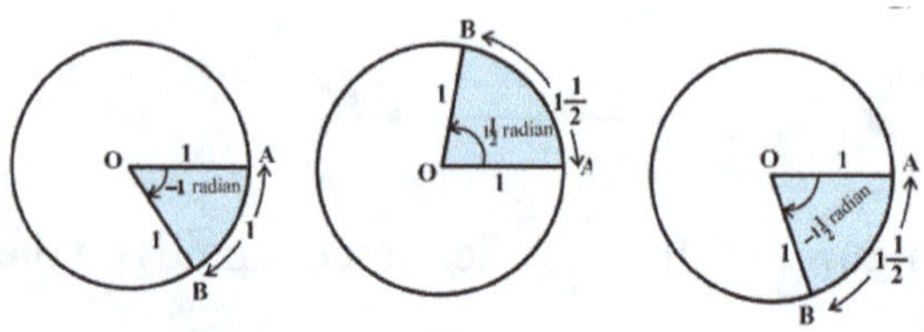

The circumference of a circle = $2\pi r$, where r is the radius of the circle. Hence, for a circle with a radius of 1 unit, the circumference is 2π. Hence, one complete revolution of the initial side subtends an angle of 2π radian at the centre. Generalizing this, we have

In a circle of radius r, an arc of length r subtends an angle of 1 radian at the centre. Hence, in a circle of radius r, an arc of length l will subtend an angle $=\frac{l}{r}$ radian. Generalizing this, we have, in a circle of radius r, if an arc of length l subtends an angle θ radian at the centre, then

$$\theta = \frac{l}{r}$$

$$\Rightarrow l = r\theta.$$

Relation between Radian and Real Numbers

Radian measures and real numbers are one and the same. Consider a unit circle with centre O. Let A be any point on the circle and OA be the initial side of the angle as shown below:

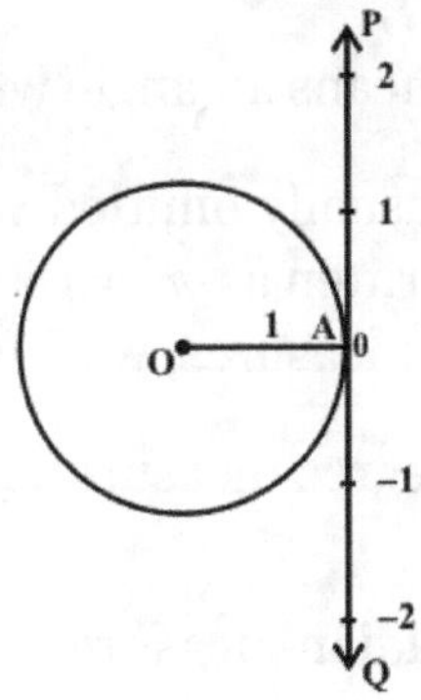

Now, consider a line PAQ drawn tangential to the circle at point A. Also, let A be the real number zero. Hence, line AP represents the positive real numbers and line AQ represents the negative real numbers. Further, let's drag the line AP along the circumference of the circle in the anticlockwise direction.

Also, let's drag the line AQ along the circumference of the circle in the clockwise direction. After doing so, we can see that every real number corresponds to a radian measure and conversely.

Relation between Degree and Radian Measures

By the definitions of degree and radian measures, we know that the angle subtended by a circle at the centre is:

- 360° – according to degree measure
- 2π radian – according to radian measure

Hence, 2π radian $=360° \Rightarrow \pi$ radian $=180°$. Now, we substitute the approximate value of π as $\frac{22}{7}$ in the equation above and get, 1 radian $=\frac{180°}{\pi}=57°16'$ approximately. Also, $1° = \frac{\pi}{180°}$ radian = 0.01746 radian approximately. Further, here is a table depicting the relationship between degree and radian measures of some common angles:

Degree	30°	45°	60°	90°	180°	270°	360°
Radian	$\frac{\pi}{6}$	$\frac{\pi}{4}$	$\frac{\pi}{3}$	$\frac{\pi}{2}$	π	$\frac{3\pi}{2}$	2π

Notational Convention

Since degree and radian measures are the two most commonly used units in angle measurement, there is a convention in place for writing them.

- If you write angle θ°, then it means an angle whose degree measure is θ.
- If you write angle β, then it means an angle whose radian measure is β.

Also, note that the term 'radian' is usually omitted while writing the radian measure. Hence, π radian $=180°$ is simply written as $\pi=180°$. Further, summing up the relationship between degree and radian measures, we have

- Radian measure $=\frac{\pi}{180°}$ x Degree measure
- Degree measure $=\frac{180°}{\pi}$ x Radian measure

Triangle

A triangle is any set of three points on a plane and the lines connecting those points to each other, as long as the three points aren't all on the same line (that would just be a line). Or, you could think of a triangle as the part of the plane that lies inside those line segments. A triangle is flat shape it has no thickness. But every triangle has a perimeter and an area, and three angles.

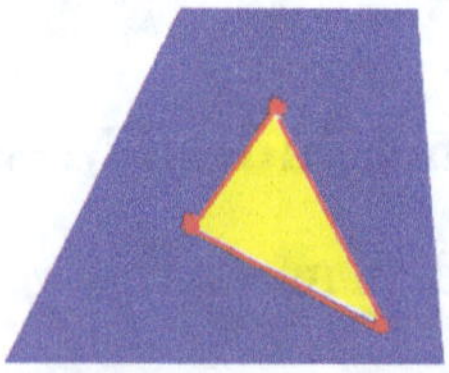

A yellow triangle lying in a blue plane

This is an equilateral triangle

The three angles of a triangle will always add up to 180 degrees, no matter how big or how small the triangle is. Think of it this way: a rectangle has four 90 degree angles, or right angles. Adding those four 90 degree angles together shows us that a rectangle has 360 degrees. But a triangle is half of a rectangle, split from corner to corner. So it has to have half the degrees of a rectangle, or 180 degrees. Because of this, if you know the measurements of two angles of a triangle, you can always figure out how big the third angle is by adding the two known angles together and subtracting that from 180 degrees.

Types of Triangles based on Sides

- Equilateral triangle: A triangle having all the three sides of equal length is an equilateral triangle.

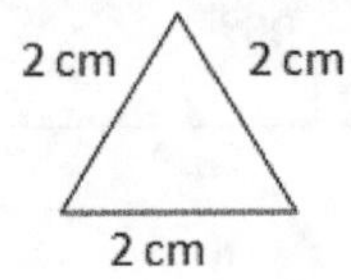

Since all sides are equal, all angles are equal too.

- Isosceles triangle: A triangle having two sides of equal length is an Isosceles triangle.

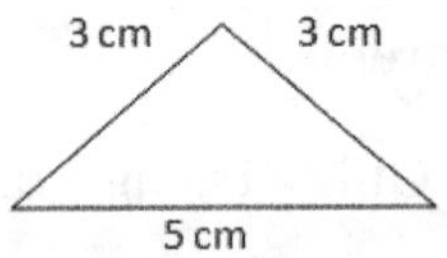

The two angles opposite to the equal sides are equal.

- Scalene triangle: A triangle having three sides of different lengths is called a scalene triangle.

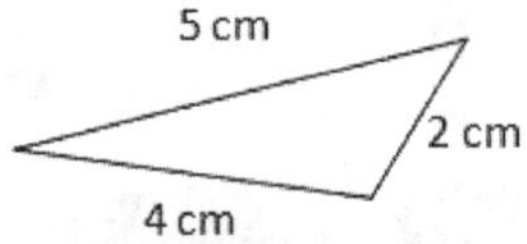

Types of Triangles based on Angles

- Acute-angled triangle: A triangle whose all angles are acute is called an acute-angled triangle or Acute triangle.

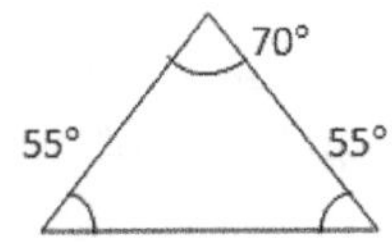

- Obtuse-angled triangle: A triangle whose one angle is obtuse is an obtuse-angled triangle or Obtuse triangle.

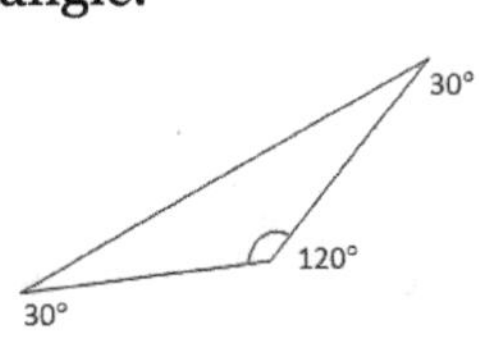

- Right-angled triangle: A triangle whose one angle is a right-angle is a Right-angled triangle or Right triangle.

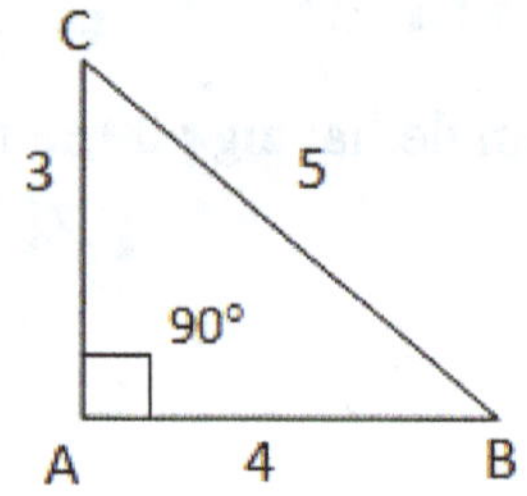

In the figure above, the side opposite to the right angle, BC is called the hypotenuse.

For a Right triangle ABC,

$$BC^2 = AB^2 + AC^2$$

This is called the Pythagorean Theorem.

In the triangle above, $5^2 = 4^2 + 3^2$. Only a triangle that satisfies this condition is a right triangle.

Hence, the Pythagorean Theorem helps to find whether a triangle is Right-angled.

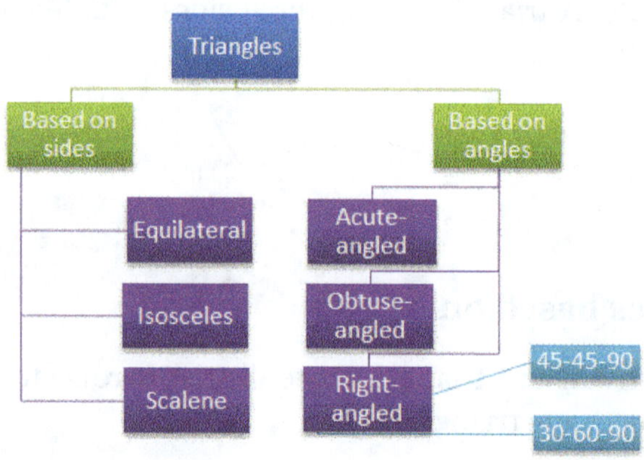

There are different types of right triangles. As of now, our focus is only on a special pair of right triangles.

1. 45-45-90 triangle
2. 30-60-90 triangle

45-45-90 triangle:

1. A 45-45-90 triangle, as the name indicates, is a right triangle in which the other two angles are 45° each.

This is an isosceles right triangle.

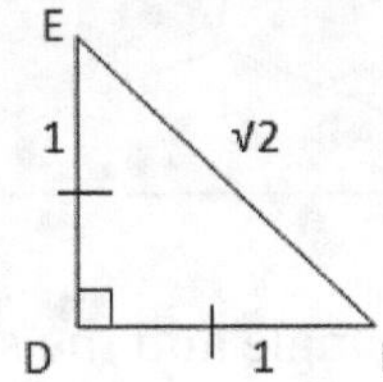

In Δ DEF, DE = DF and ∠D = 90°.

The sides in a 45-45-90 triangle are in the ratio $1:1:\sqrt{2}$.

30-60-90 triangle:

1. A 30-60-90 triangle, as the name indicates, is a right triangle in which the other two angles are 30° and 60°.

This is a scalene right triangle as none of the sides or angles are equal.

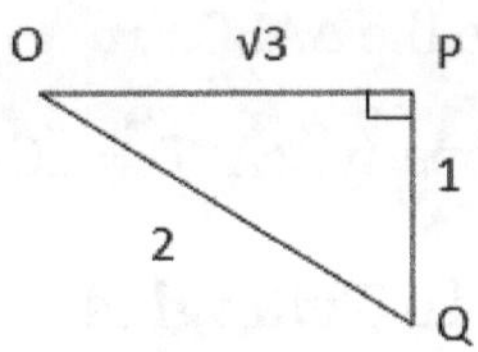

The sides in a 30-60-90 triangle are in the ratio $1:\sqrt{3}:2$

Like any other right triangle, these two triangles satisfy the Pythagorean Theorem.

Basic Properties of Triangles

- The sum of the angles in a triangle is 180°. This is called the angle-sum property.
- The sum of the lengths of any two sides of a triangle is greater than the length of the third side. Similarly, the difference between the lengths of any two sides of a triangle is less than the length of the third side.
- The side opposite to the largest angle is the longest side of the triangle and the side opposite to the smallest angle is the shortest side of the triangle.

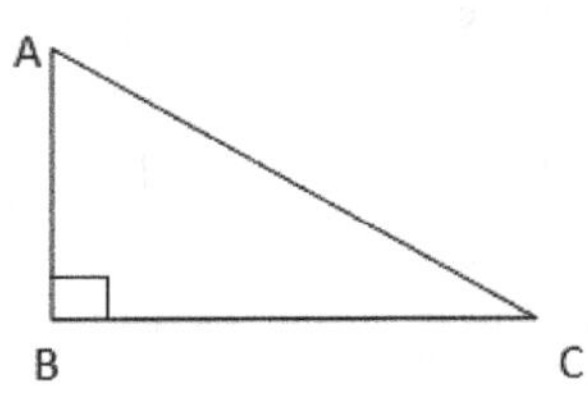

In the figure above, ∠B is the largest angle and the side opposite to it (hypotenuse), is the largest side of the triangle.

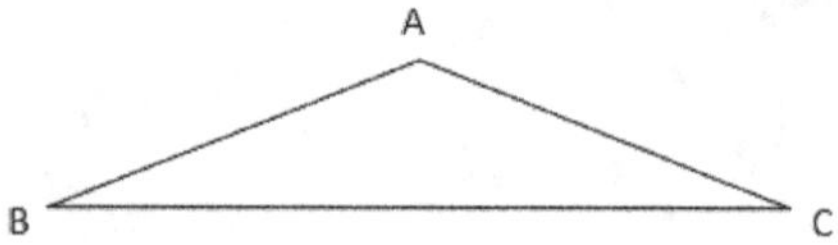

In the figure above, ∠A is the largest angle and the side opposite to it, BC is the largest side of the triangle.

- An exterior angle of a triangle is equal to the sum of its interior opposite angles. This is called the exterior angle property of a triangle.

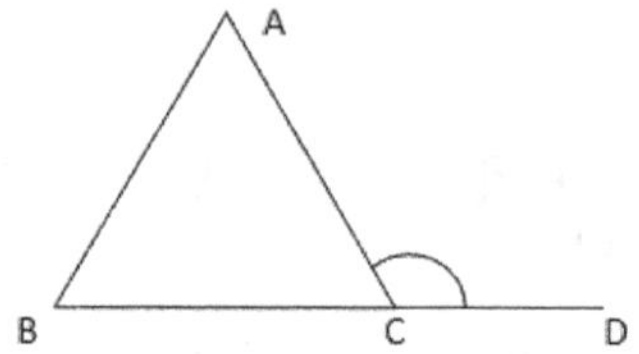

Here, ∠ACD is the exterior angle to the ΔABC.

According to the exterior angle property, ∠ACD = ∠CAB + ∠ABC.

Similarity and Congruency in Triangles

Figures with same size and shape are congruent figures. If two shapes are congruent, they remain congruent even if they are moved or rotated. The shapes would also remain congruent if we reflect the shapes by producing mirror images. Two geometrical shapes are congruent if they cover each other exactly.

Figures with same shape but with proportional sizes are similar figures. They remain similar even if they are moved or rotated.

Similarity of Triangles

Two triangles are said to be similar if the corresponding angles of two triangles are congruent and lengths of corresponding sides are proportional.

It is written as Δ ABC ~ Δ XYZ and said as Δ ABC 'is similar to' Δ XYZ.

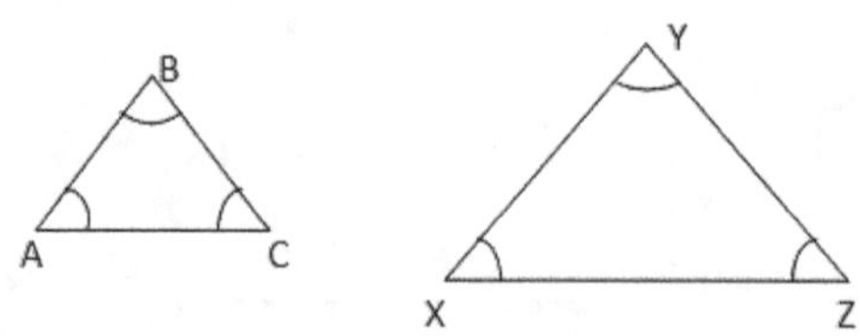

Here, ∠A = ∠X, ∠B = ∠Y and ∠C = ∠Z AND

AB / XY = BC / YZ = CA / ZX

The necessary and sufficient conditions for two triangles to be similar are as follows:

(1) Side-Side-Side (SSS) criterion for similarity:

If three sides of a triangle are proportional to the corresponding three sides of another triangle then the triangles are said to be similar.

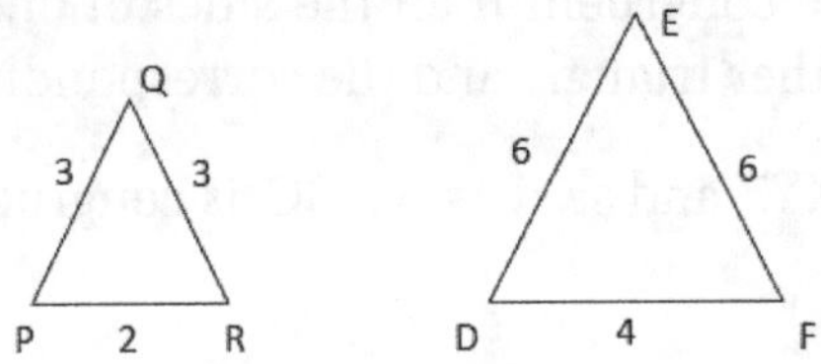

Here, Δ PQR ~ Δ DEF as:

PQ / DE = QR / EF = RP / FD

(2) Side-Angle-Side (SAS) criterion for similarity:

If the corresponding two sides of the two triangles are proportional and one included angle is equal to the corresponding included angle of another triangle then the triangles are similar.

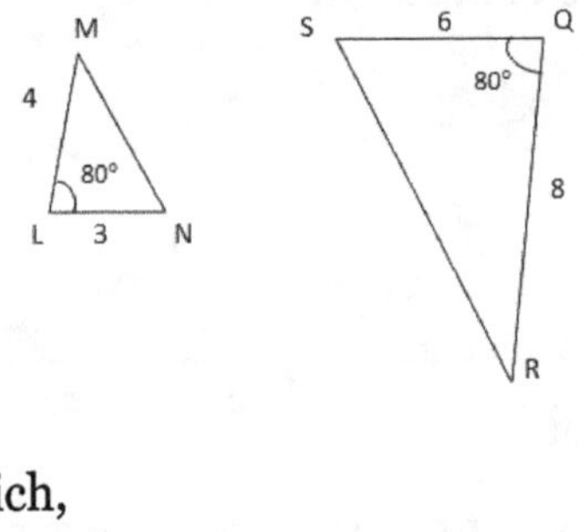

Here, Δ LMN ~ Δ QRS in which,

∠L = ∠Q

QS / LN = QR / LM

(3) Angle-Angle-Angle (AAA) criterion for similarity:

If the three corresponding angles of the two triangles are equal then the two triangles are similar.

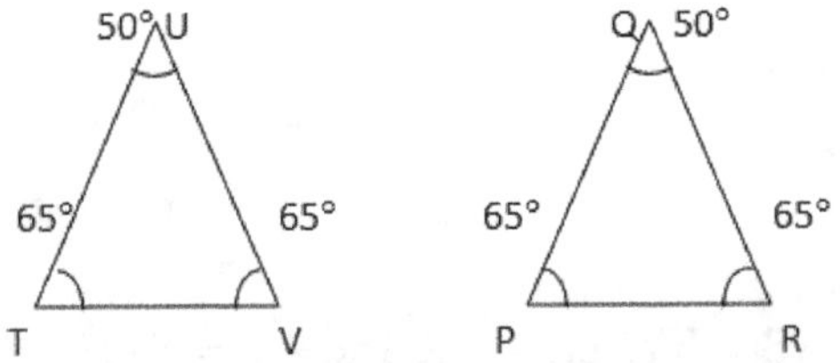

Here Δ TUV ~ Δ PQR as

$\angle T = \angle P$, $\angle U = \angle Q$ and $\angle V = \angle R$

Congruency of Triangles

Two triangles are said to be congruent if all the sides of one triangle are equal to the corresponding sides of another triangle and the corresponding angles are equal.

It is written as Δ ABC ≅ Δ XYZ and said as Δ ABC 'is congruent to' Δ XYZ.

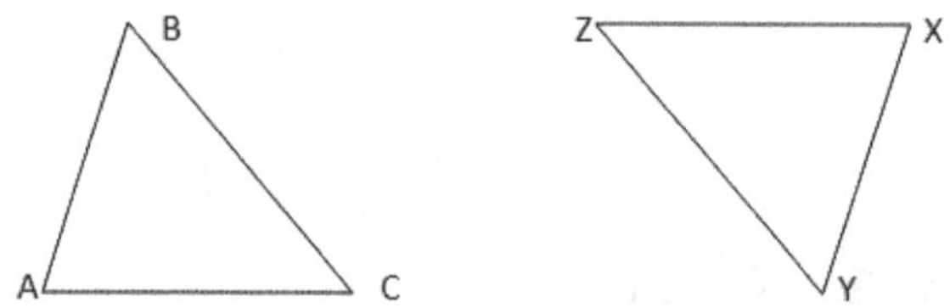

The necessary and sufficient conditions for two triangles to be congruent are as follows:

(1) Side-Side-Side (SSS) criterion for congruence:

If three sides of a triangle are equal to the corresponding three sides of another triangle then the triangles are said to be congruent.

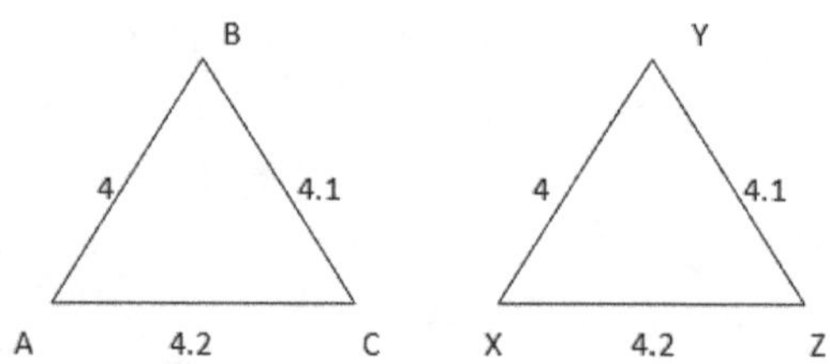

Here, Δ ABC ≅ Δ XYZ as AB = XY, BC = YZ and AC = XZ.

(2) Side-Angle-Side (SAS) criterion for congruence:

If two sides and the angle included between the two sides of a triangle are equal to the corresponding two sides and the included angle of another triangle, then the triangles are congruent.

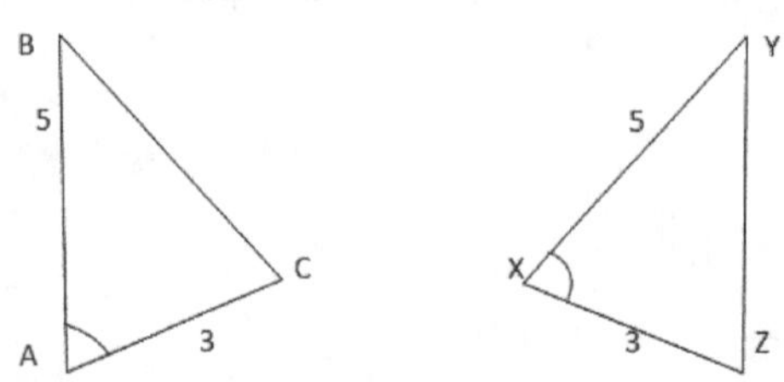

Here, Δ ABC ≅ Δ XYZ as AB = XY, $\angle A = \angle X$ and AC = XZ.

(3) Angle-Side-Angle (ASA) criterion for congruence:

If two angles and the included side of a triangle are equal to the corresponding two angles and the included side of another triangle then the triangles are congruent.

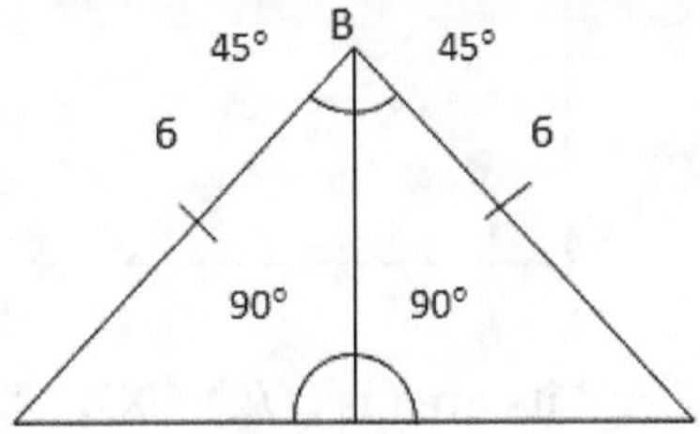

In the figure above, Δ ABD ≅ Δ CBD in which,

∠ABD = ∠CBD, AB = CB and ∠ADB = ∠CDB.

(4) Right-Angle Hypotenuse criterion of congruence:

If the hypotenuse and one side of a right-angled triangle are equal to the corresponding hypotenuse and side of another right-angled triangle, then the triangles are congruent.

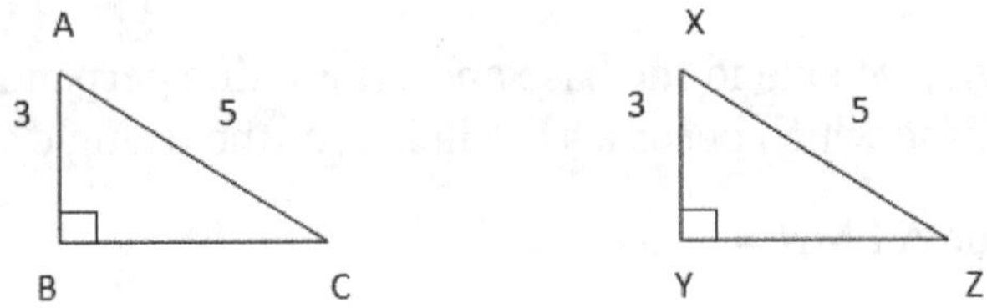

Here, ∠B = ∠Y = 90° and AB = XY, AC = XZ.

Area of a Triangle

The Area of a triangle is given by the formula:

Area of a triangle = (1/2) *Base * Height

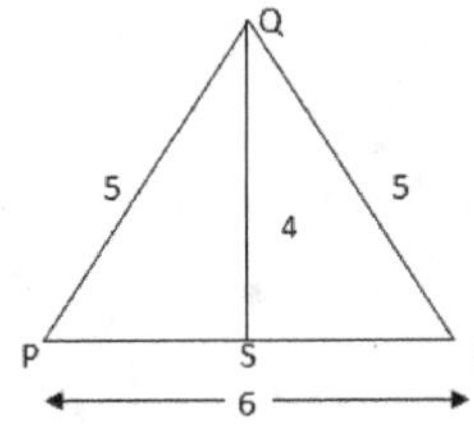

To find the area of a triangle, we draw a perpendicular line from the base to the opposite vertex which gives the height of the triangle.

So the area of the Δ PQR = (1/2) * (PR * QS) = (1/2) * 6 *4 =12 sq. units.

For a right triangle, it's easy to find the area as there is a side perpendicular to the base, so we can consider it as height.

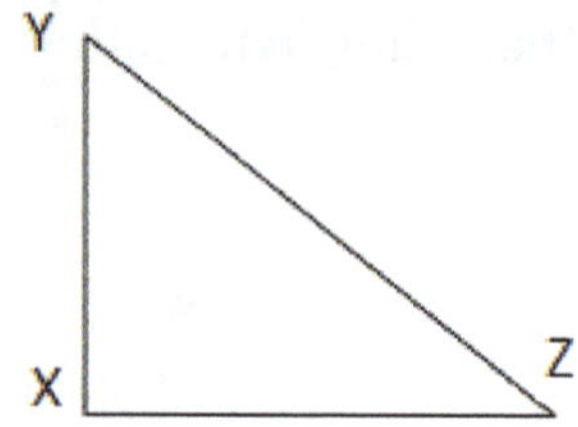

The height of the Δ XYZ is XY and its area is (1/2) * XZ * XY sq. units.

Now, how do we find the area of an obtuse triangle LMN ?

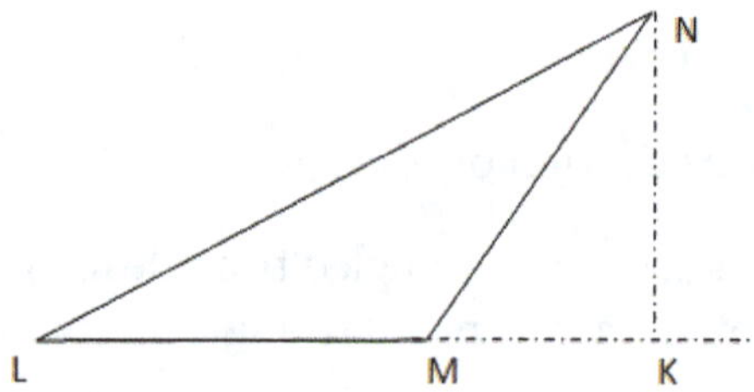

For an obtuse triangle, we extend the base and draw a line perpendicular from the vertex to the extended base which becomes the height of the triangle.

Hence, the area of the Δ LMN = (1/2) * LM * NK sq. units.

Circle

A circle is a round shaped figure that has no corners or edges.

In geometry, a circle can be defined as a closed, two-dimensional curved shapes.

Center of a Circle

The center of a circle is the center point in a circle from which all the distances to the points on the circle are equal. This distance is called the radius of the circle.

Here, point P is the center of the circle.

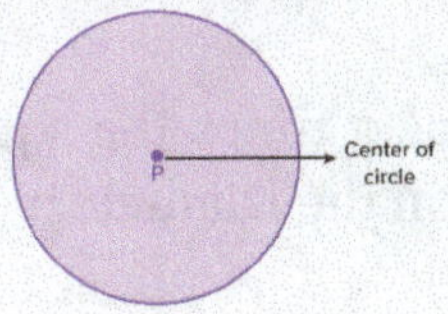

Semicircle

A semi-circle is half of a circle, formed by cutting a whole circle along a line segment passing through the center of the circle. This line segment is called the diameter of the circle.

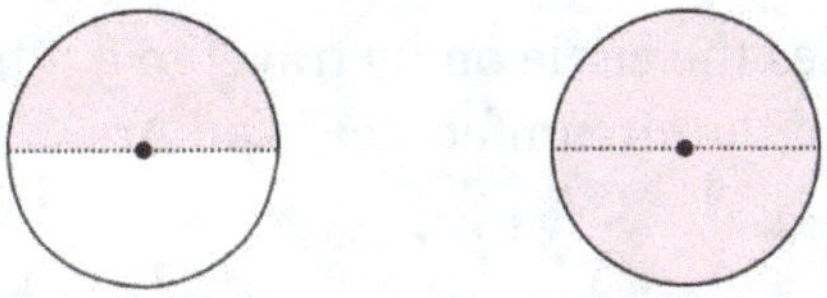
Half circle or Semicircle 2 Semicircles in 1 circle

Quarter Circle

A quarter circle is a quarter of a circle, formed by splitting a circle into 4 equal parts or a semicircle into 2 equal parts.

A quarter circle is also called a quadrant.

Quarter or Quadrant 4 Quarters in a circle

Important Terms Related to Circle

Diameter

The diameter can be termed as a line which is drawn across a circle passing through the center.

Radius

The distance from the middle or center of a circle towards any point on it is a radius. Interestingly, when you place two radii back-to-back, the resultant would hold the same length as one diameter. Therefore, we can call one diameter twice as long as the concerned radius.

Area of Circle

In a circle, the area can be stated as π times the square of the radius. It is written as: $A = \pi r^2$. Taking into consideration the Diameter: $A = (\pi / 4) \times D^2$

Chord

A line segment that joins two points present on a curve is called as the chord. In geometry, the usefulness of a chord is focused on describing a line segment connecting two endpoints which rest on a circle.

Tangent and Arc

A line which slightly touches the circle on its travel to a different direction is Tangent. On the other hand, a part of the circumference is an Arc.

Sector and Segment

A sector is a part of a circle surrounded by two radii of it together with their intercepted arc. The segment is that region which is enclosed by a chord together with the arc subtended by the chord.

Common Sectors

In geometry, Quadrant and Semicircle are known as two special versions of a sector.

- A circle's quarter is termed as Quadrant.
- Half a circle is known as a Semicircle.

Properties and Key Aspects

Focusing on geometry, there are numerous facts associated with circles. Further, the relation of it to straight lines, polygons, and angles can also be proved. All of these facts together are properties of the circle. Let us try to learn the primary properties in order to enhance our knowledge.

- Circles holding equal radii are known to be congruent.
- To your surprise, circles with different radii are seen as similar.
- In a circle, the central angle that intercepts an arc is known to be double to any inscribed angle which intercepts the same arc.
- The chords that are equidistant from the center are known to be of the same length.

- A radius perpendicular to a particular chord does bisect the chord.
- The tangent is always at right angles to the radius considering the point of contact.
- Two tangents which are drawn on a circle from an exterior point are equal in length.
- The circumference of two diverse circles is proportional to the corresponding radii.
- The angle subtended at the circle's center by its circumference is known to be equivalent to four right angles.
- Arcs associated to the same circle are termed proportional to their corresponding angles.
- Equal chords hold equal circumferences.
- Equal circles hold equal circumferences.
- Radii linked to the same or equal circles are known to be equal.
- The longest chord is the diameter.

Square

In plane (Euclidean) geometry, a square is a regular polygon with four sides. It may also be thought of as a special case of a rectangle, as it has four right angles and parallel sides. Likewise, it is also a special case of a rhombus, kite, parallelogram, and trapezoid.

Mensuration Formulae

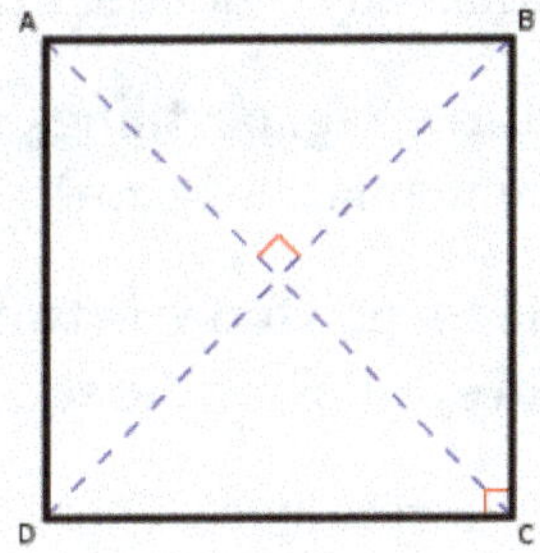

A Square

The area of a square is the product of the length of its sides.

$$P = 4t.$$

And the area is

$$A = t^2.$$

In classical times, the second power was described in terms of the area of a square, as in the above formula. This led to the use of the term square to mean raising to the second power.

Standard Coordinates

The coordinates for the vertices of a square centered at the origin and with side length 2 are $(\pm 1,\ \pm 1)$, while the interior of the same consists of all points (x_0, x_1) with $-1 < x_i < 1$.

Properties

Each angle in a square is equal to 90 degrees, or a right angle.

The diagonals of a square are equal. Conversely, if the diagonals of a rhombus are equal, then that rhombus must be a square. The diagonals of a square are $\sqrt{2}$ (about 1.41) times the length of a side of the square. This value, known as Pythagoras' constant, was the first number proven to be irrational.

If a figure is both a rectangle (right angles) and a rhombus (equal edge lengths) then it is a square.

Other Facts

- If a circle is circumscribed around a square, the area of the circle is $\pi / 2$ (about 1.57) times the area of the square.
- If a circle is inscribed in the square, the area of the circle is $\pi / 4$ (about 0.79) times the area of the square.
- A square has a larger area than any other quadrilateral with the same perimeter.
- A square tiling is one of three regular tilings of the plane (the others are the equilateral triangle and the regular hexagon).
- The square is in two families of polytopes in two dimensions: hypercube and the cross polytope. The Schläfli symbol for the square is {4}.
- The square is a highly symmetric object. There are four lines of reflectional symmetry and it has rotational symmetry through 90°, 180° and 270°. Its symmetry group is the dihedral group D_4.
- If the area of a given square with side length S is multiplied by the area of a "unit triangle" (an equilateral triangle with side length of 1 unit), which is $\frac{\sqrt{}}{}$ units squared, the new area is that of the equilateral triangle with side length S.

Non-Euclidean Geometry

In non-euclidean geometry, squares are more generally polygons with four equal sides and equal angles.

In spherical geometry, a square is a polygon whose edges are great circle arcs of equal distance, which meet at equal angles. Unlike the square of plane geometry, the angles of such a square are larger than a right angle.

In hyperbolic geometry, squares with right angles do not exist. Rather, squares in hyperbolic geometry have angles of less than right angles. Larger squares have smaller angles.

Examples:

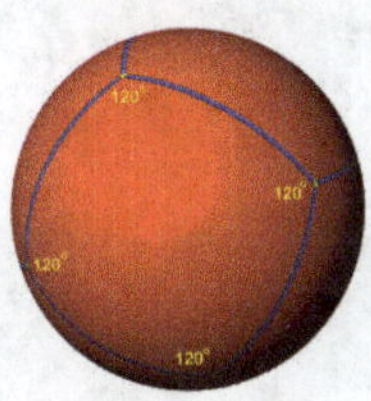

Six squares can tile the sphere with three squares around each vertex and 120 degree internal angles. This is called a spherical cube. The Schläfli symbol is {4,3}.

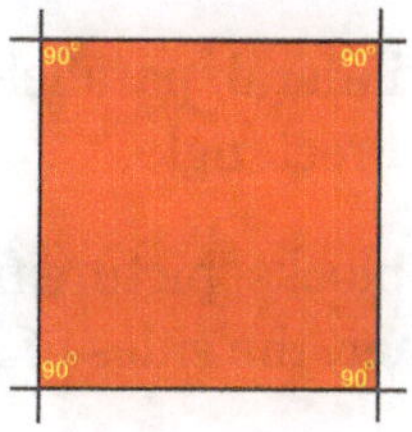

Squares can tile the Euclidean plane with four around each vertex, with each square having an internal angle of 90 degrees. The Schläfli symbol is {4,4}.

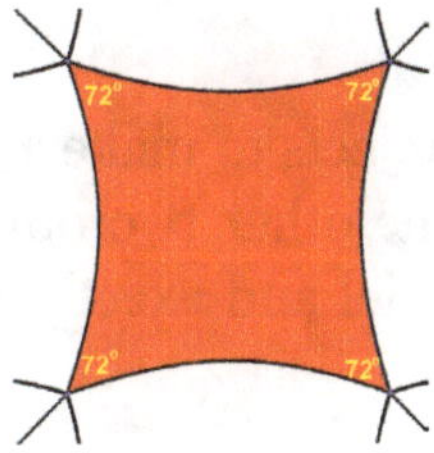

Squares can tile the hyperbolic plane with five around each vertex, with each square having 72 degree internal angles. The Schläfli symbol is {4,5}.

Rectangle

For a shape to be a rectangle, it must be a four-sided polygon with two pairs of parallel, congruent sides and four interior angles of 90° each. If you have a shape that matches that description, it also is all this:

- A plane figure
- A closed shape
- A quadrilateral
- A parallelogram

The four sides of your polygon, to create two pairs of parallel sides, must also be two congruent pairs. The base and top will be equal in length, and the left and right sides will be equal in length.

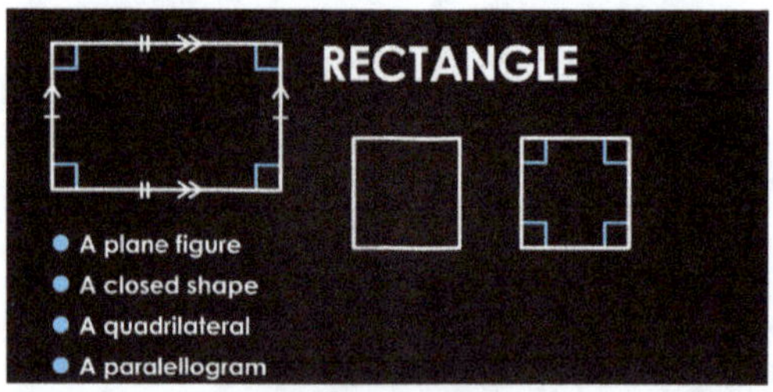

Relation between Square and Rectangle

While a rectangle is a type of quadrilateral, parallelogram, closed shape and plane figure, only a square is always a type of rectangle.

You can see them in bricks, cement blocks, picture frames, posters, sheets of paper, the faces of play bricks that snap together, the sides of shoe boxes and cereal boxes, and a lot of other everyday objects.

Rectangles are great because they stack neatly, since they have two pairs of parallel sides. Their right angles make sure built things (houses, office buildings, schools) stand straight and tall.

You can use four linear (straight) objects to make a rectangle. Make certain two of the objects are the same length, and the other two objects are the same length. Arrange them so the longer pieces are parallel and exactly a distance apart so the other two, shorter pieces can touch their ends.

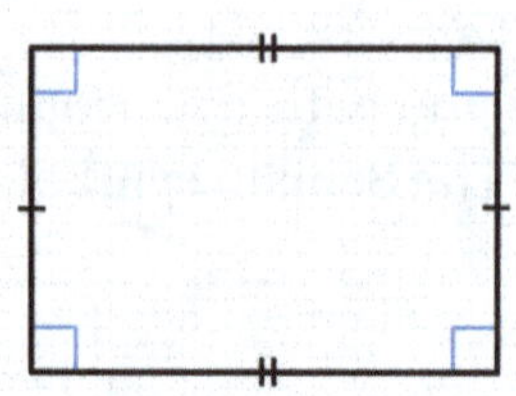

Rectangle

When all four ends are touching, you may need to adjust to make sure all four inside angles look like right angles, or 90°. Adjacent sides of a rectangle are perpendicular.

Properties of Rectangles

The main identifying property of a rectangle is its four interior right angles. You cannot construct a rectangle without those four angles adding to 360° and each measuring 90°. When you do that, the four sides will automatically create the other identifying property.

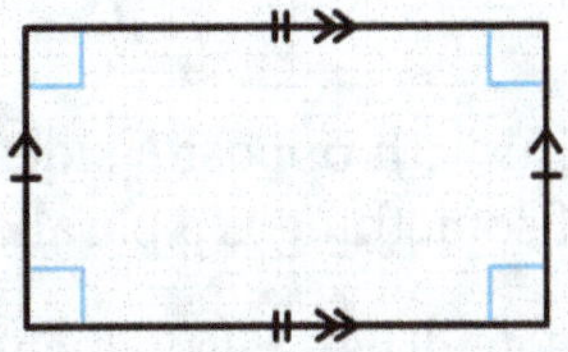

Identifies of a properties rectangles, its four interior right angles

The other property that identifies rectangles is that opposite sides are congruent and parallel. Congruent means they have the same length; parallel means they are the same distance apart throughout their length.

Construction of a Rectangle

A protractor measures angles. A straightedge (or ruler) makes straight lines. Use the ruler or straightedge to make a straight line segment on a piece of paper, roughly in the lower third of the sheet. That line segment is your base. Align your protractor with that line segment, working at one endpoint at a time.

At each end of the line segment, exactly at the endpoint, mark a 90° angle. Use the straightedge to connect that 90° mark and the endpoint of the line segment. You now have the two sides of the rectangle.

Mark a new endpoint on one of those new sides, at some distance away from your base. Turn the protractor 90° and align it with either side, at the endpoint of that newly drawn side. Mark 90° using the protractor and again use the straightedge to connect the side's endpoint with that new 90° mark. When that line segment intersects the other side, you have constructed a rectangle.

If you use a ruler with markings in inches or centimeters, you can measure the length

of your base, measure the length of the two sides left and right, and ensure the long, top side is equal to the length of the base.

Parallelogram

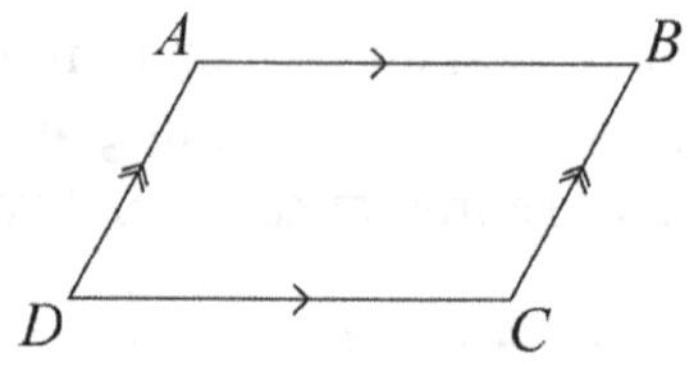

A parallelogram is a quadrilateral with opposite sides parallel. But there are various tests that can be applied to see if something is a parallelogram.

It is the "parent" of some other quadrilaterals, which are obtained by adding restrictions of various kinds:

- A rectangle is a parallelogram but with all four interior angles fixed at 90°.
- A rhombus is a parallelogram but with all four sides equal in length.
- A square is a parallelogram but with all sides equal in length and all interior angles 90°.

A quadrilateral is a parallelogram if:

1. Both pairs of opposite sides are parallel. (By definition).
2. Both pairs of opposite sides are congruent. If they are congruent, they must also be parallel.
3. One pair of opposite sides are congruent *and* parallel. Then, the other pair must also be parallel.

Properties of a Parallelogram

These facts and properties are true for parallelograms and the descendant shapes: square, rectangle and rhombus.

Base

Any side can be considered a base. Choose any one you like. If used to calculate the area the corresponding altitude must be used. In the figure above, one of the four possible bases and its corresponding altitude has been chosen.

Altitude (Height)

The altitude (or height) of a parallelogram is the perpendicular distance from the base to the opposite side (which may have to be extended). In the figure above, the altitude corresponding to the base CD is shown.

Area

The area of a parallelogram can be found by multiplying a base by the corresponding altitude.

Perimeter

The distance around the parallelogram. The sum of its sides.

Opposite Sides

Opposite sides are congruent (equal in length) and parallel.

Diagonals

Each diagonal cuts the other diagonal into two equal parts, as in the diagram below.

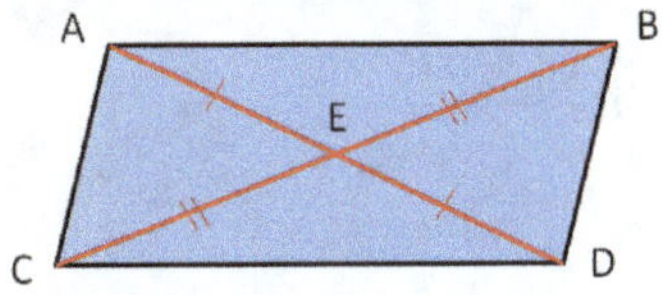

Interior Angles

1. Opposite angles are equal as can be seen below.
2. Consecutive angles are always supplementary (add to 180°).

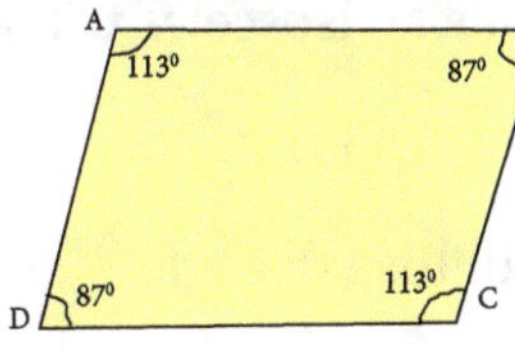

Parallelogram Inscribed in any Quadrilateral

If you find the midpoints of each side of any quadrilateral, then link them sequentially with lines, the result is always a parallelogram.

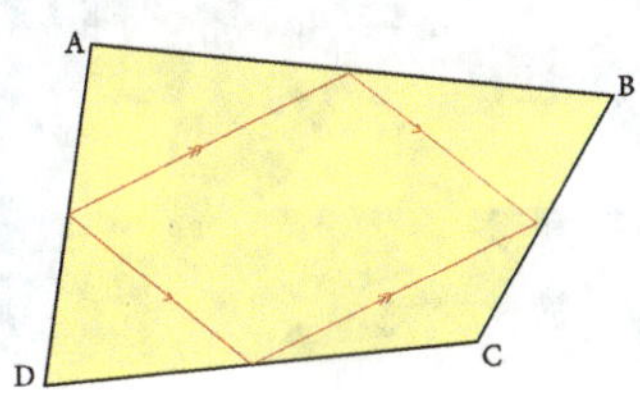

Trapezoid

A trapezium is a quadrilateral wherein one pair of the opposite sides are parallel while the other isn't.

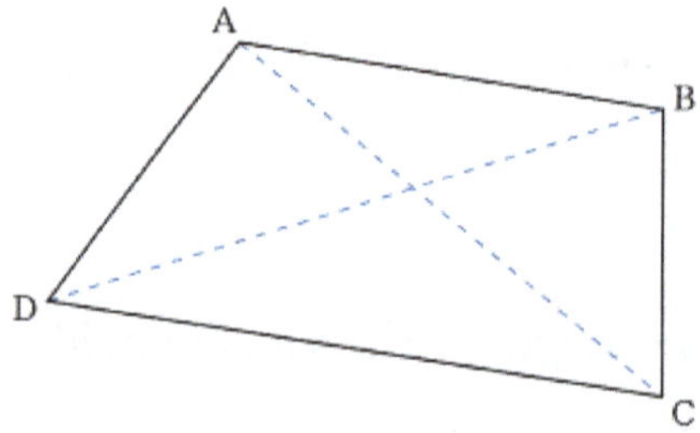

Suppose if ABCD is a trapezium, so we say that AB||CD but yes AD and BC are not parallel. The trapezium is basically a triangle with the top sliced off. A trapezium that has equal non-parallel sides and equal base angles is an isosceles trapezium.

The parallel sides of this quadrilateral are the base and to calculate the height you need to draw a perpendicular from one parallel side to other.

$$\text{Area of a Trapezium } = h\frac{(a+b)}{2}$$

Types of Trapezoids

Since trapezoids can begin life as triangles, they share names derived from the kinds of triangles:

1. Scalene trapezoid - Started out as a scalene triangle
2. Isosceles trapezoid - Began as an isosceles triangle
3. Right trapezoid - Once was a right triangle
4. Obtuse trapezoid - Like an obtuse triangle
5. Acute trapezoid - Like an acute triangle

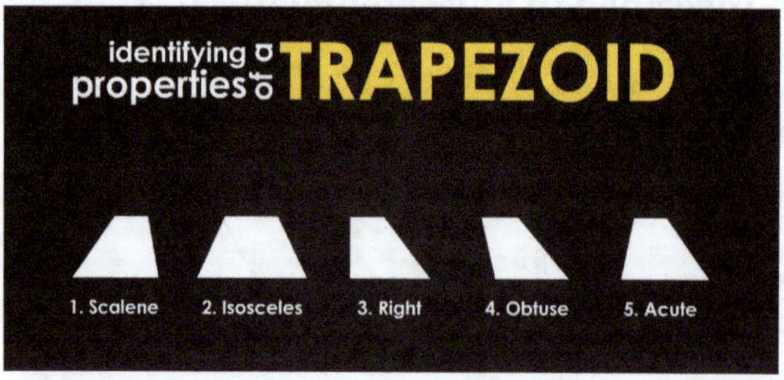

Scalene Trapezoid

A scalene trapezoid has four sides of unequal length. The bases are parallel but of different lengths. The two legs are of different lengths.

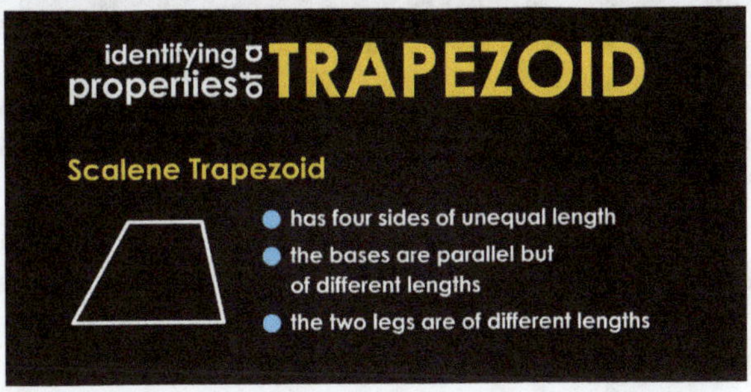

Isosceles Trapezoid

An isosceles trapezoid has legs of equal length. The bases are parallel but of different lengths.

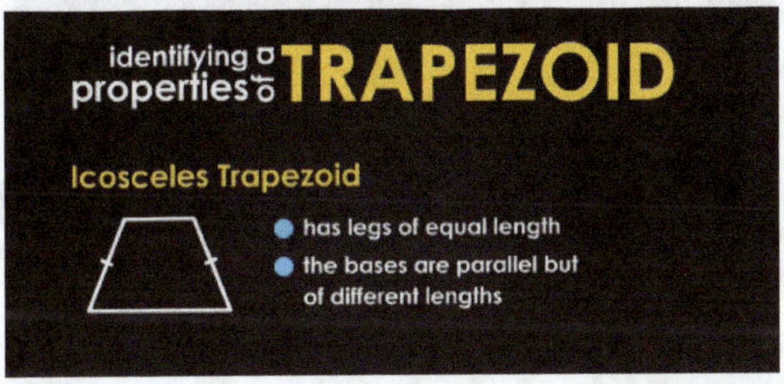

Right Trapezoid

A right trapezoid has one right angle (90°) between either base and a leg.

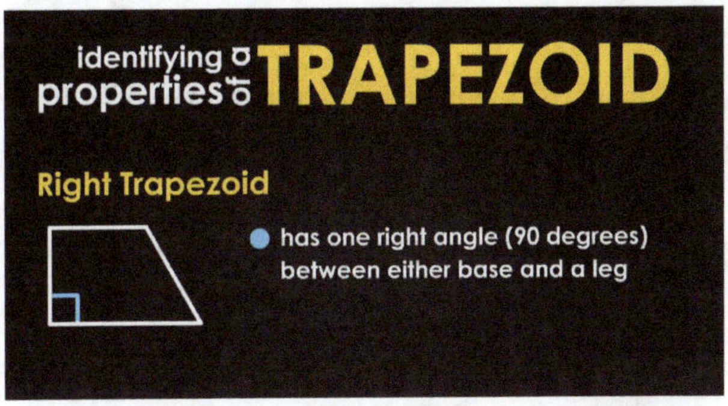

Obtuse Trapezoid

An obtuse trapezoid has one interior angle (created by either base and a leg) greater than 90°.

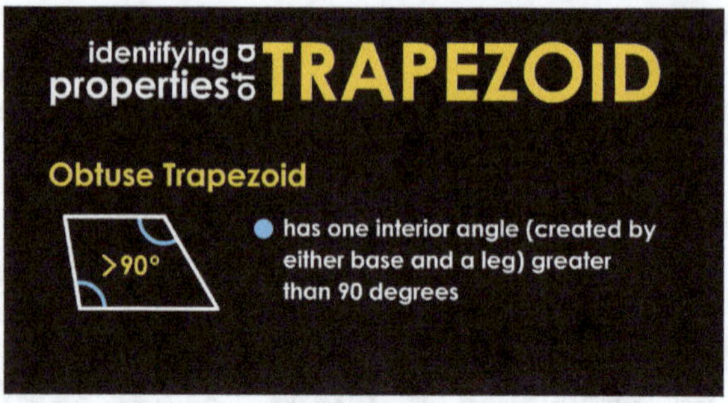

Acute Trapezoid

An acute trapezoid has both interior angles (created by the longer base and legs) measuring less than 90°.

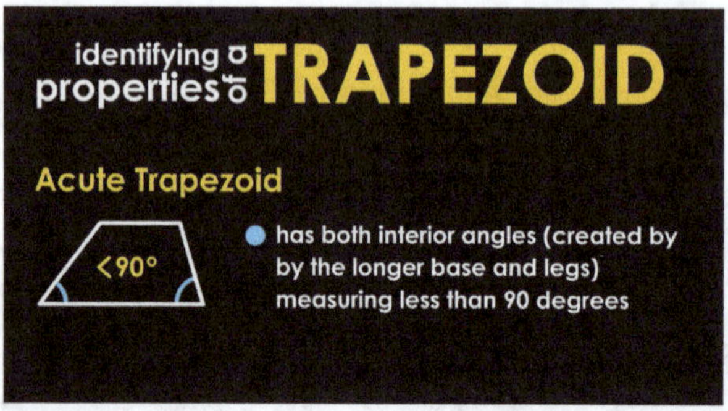

Solid Geometry

Solid geometry is concerned with three-dimensional shapes. Some examples of three-dimensional shapes are cubes, rectangular solids, prisms, cylinders, spheres, cones and pyramids.

The following figures show some examples of shapes in solid geometry.

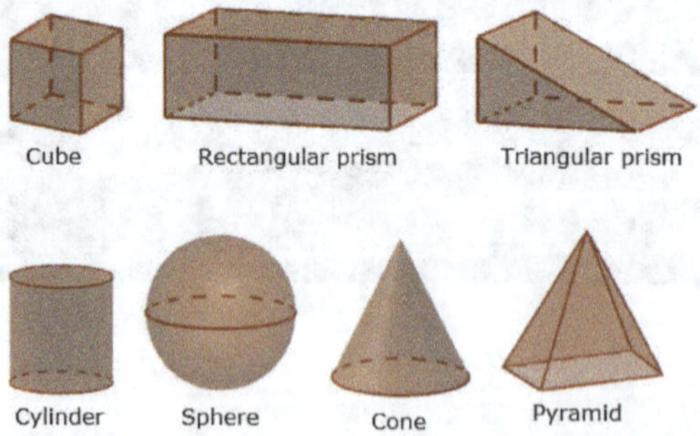

The following table gives the volume formulas and surface area formulas for the following solid shapes: Cube, Rectangular Prism, Prism, Cylinder, Sphere, Cone, and Pyramid.

Shape	Volume	Surface Area
Cube With side length, s	s^3	$6s^2$
Rectangular Prim With length, l, width, w and height, h	lwh	$2(lw+lh+wh)$
Prism With area of base, B, perimeter of base, P, and height, h	Bh	$2B+Ph$
Cylinder With radius, r and height, h	$\pi r^2 h$	$\pi r^2+2\pi rh$
Sphere With radius, r	$\frac{4}{3}\pi r^3$	$4\pi r^2$
Cone With radius, r, vertical height, h, and slant height, s	$\frac{1}{3}\pi r^2 h$	$\pi r^2+\pi rs$
Pyramid With area of base, B, perimeter of base, P, vertical height, h and slant height, s	$\frac{1}{3}Bh$	Regular Pyramid $B+\frac{1}{2}Ps$

Cubes

A cube is a three-dimensional figure with six matching square sides.

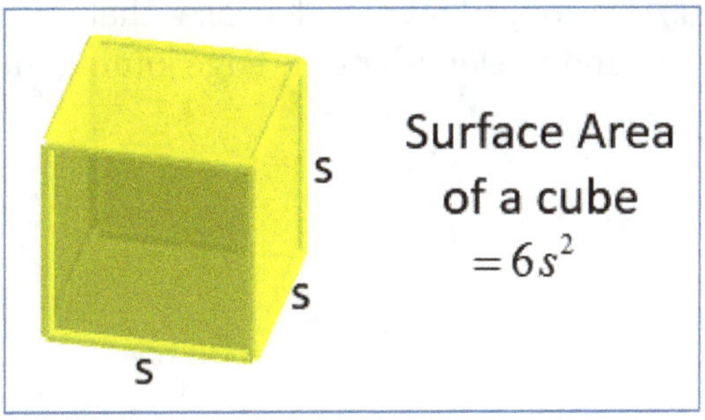

The figure above shows a cube. The dotted lines indicate edges hidden from your view.

If *s* is the length of one of its sides, then the volume of the cube is $s \times s \times s$

Volume of the cube = s^3

The area of each side of a cube is s^2. Since a cube has six square-shape sides, its total surface area is 6 times s^2.

Surface area of a cube = $6s^2$.

Rectangular Prisms or Cuboids

A rectangular prism is also called a rectangular solid or a cuboid.

In a rectangular prism, the length, width and height may be of different lengths.

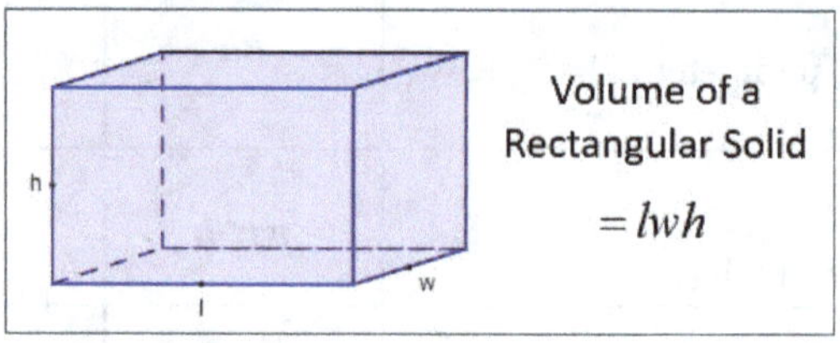

The volume of the above rectangular prism would be the product of the length, width and height that is:

Volume of rectangular prism = lwh

Total area of top and bottom surfaces is $lw + lw = 2lw$

Total area of front and back surfaces is $lh + lh = 2lh$

Total area of the two side surfaces is $wh + wh = 2wh$

Surface area of rectangular prism = $2lw + 2lh + 2wh = 2(lw + lh + wh)$.

Prisms

A prism is a solid that has two congruent parallel bases that are polygons. The polygons form the bases of the prism and the length of the edge joining the two bases is called the height.

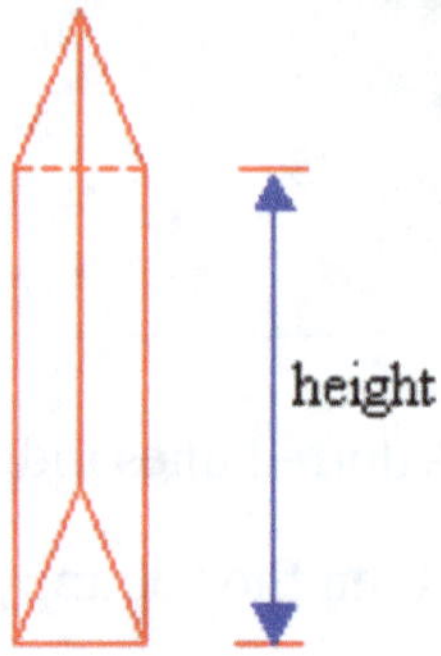

Triangle-shaped base

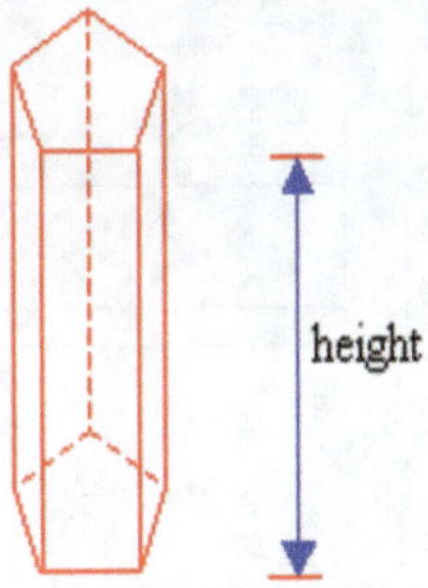

Pentagon-shaped base

The above diagrams show two prisms: one with a triangle-shaped base called a triangular prism and another with a pentagon-shaped base called a pentagonal prism.

A rectangular solid is a prism with a rectangle-shaped base and can be called a rectangular prism.

The volume of a prism is given by the product of the area of its base and its height.

Volume of prism = area of base × height

The surface area of a prism is equal to 2 times area of base plus perimeter of base times height.

Surface area of prism = 2 × area of base + perimeter of base × height

Cylinders

A cylinder is a solid with two congruent circles joined by a curved surface.

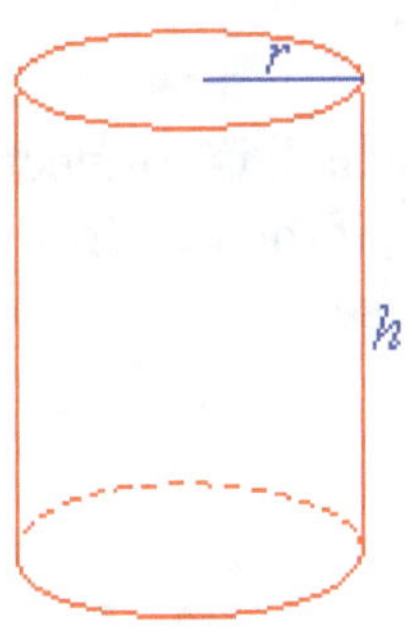

In the above figure, the radius of the circular base is r and the height is h. The volume of the cylinder is the area of the base × height.

Volume of cylinder $\pi r^2 h$

The net of a solid cylinder consists of 2 circles and one rectangle. The curved surface opens up to form a rectangle.

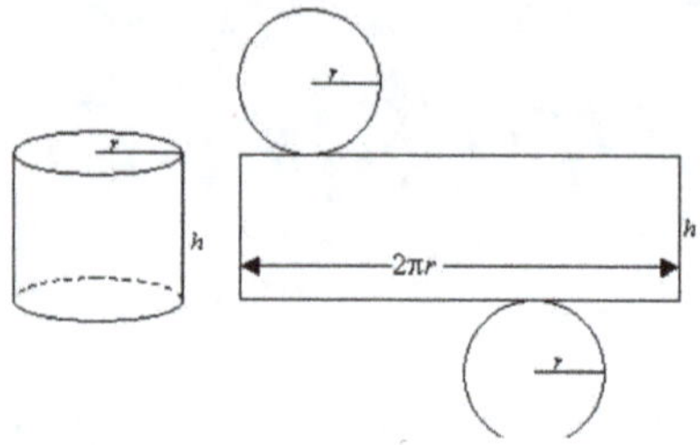

Surface area = 2 × area of circle + area of rectangle

Surface area of cylinder $= 2\pi r^2 + 2\pi rh = 2\pi r(r + h)$

Spheres

A sphere is a solid with all its points the same distance from the center.

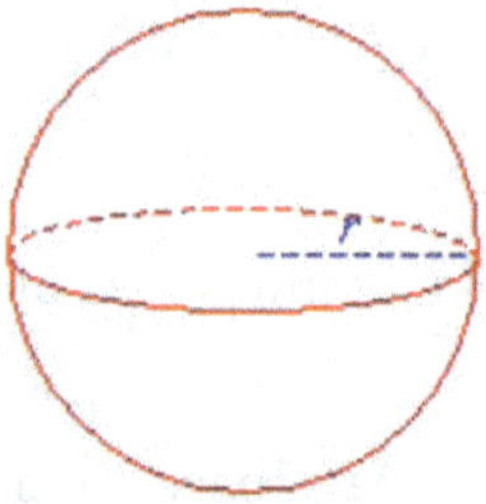

Volume of sphere $= \frac{4}{3}\pi r^3$

Surface area of sphere $= 4\pi r^2$

Cones

A circular cone has a circular base, which is connected by a curved surface to its vertex. A cone is called a right circular cone, if the line from the vertex of the cone to the center of its base is perpendicular to the base.

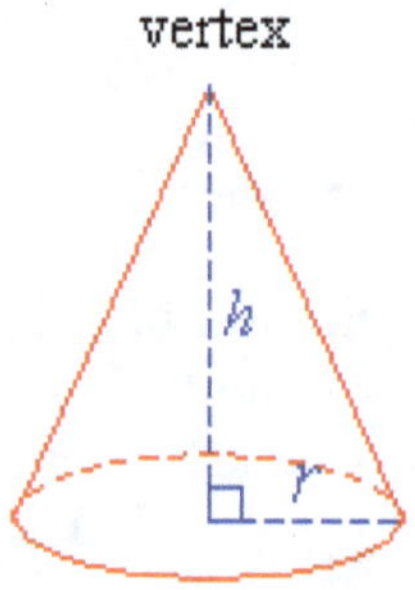

Volume of cone $\frac{1}{3}\pi r^2 h$

The net of a solid cone consists of a small circle and a sector of a larger circle. The arc of the sector has the same length as the circumference of the smaller circle.

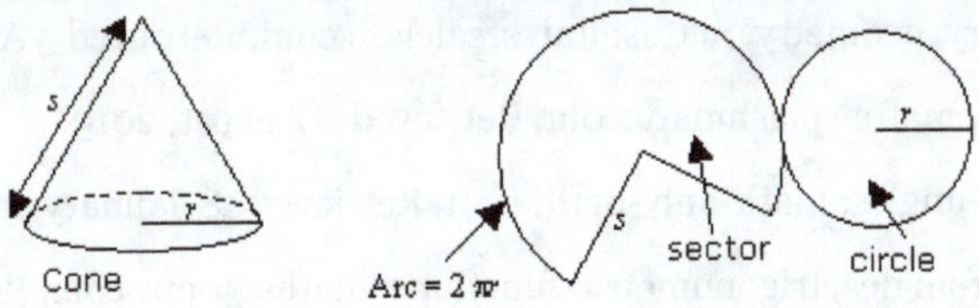

Surface area of cone = Area of sector + area of circle

$= \pi rs + \pi r^2 = \pi r(r + s)$

Pyramids

A pyramid is a solid with a polygon base and connected by triangular faces to its vertex. A pyramid is a regular pyramid if its base is a regular polygon and the triangular faces are all congruent isosceles triangles.

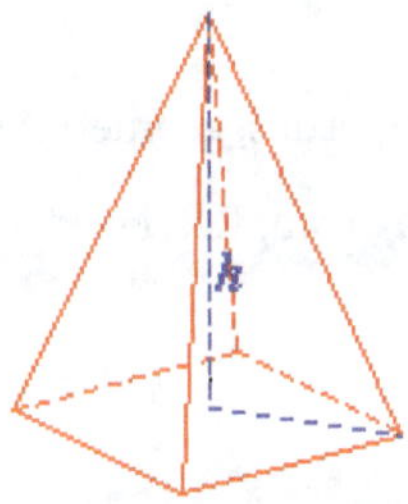

Volume of pyramid $= \frac{1}{3} \times$ area of base $\times$ height

Nets of a Solid

An area of study closely related to solid geometry is nets of a solid. Imagine making cuts along some edges of a solid and opening it up to form a plane figure. The plane figure is called the net of the solid.

The following figures show the two possible nets for the cube.

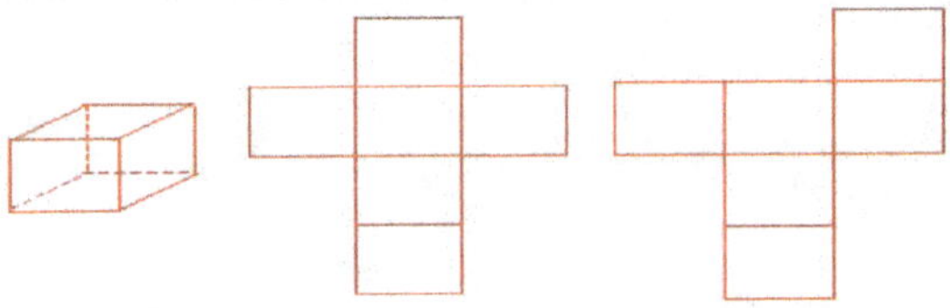

References

- What-is-geometry: thoughtco.com, Retrieved 25 August, 2019
- What-is-geometry, geometry, maths: universalclass.com, Retrieved 5 April, 2019
- Angle, geometry, math: splashmath.com, Retrieved 8 August, 2019
- Classification-of-angles: math-only-math.com, Retrieved 18 January, 2019
- Measurement-of-angles, trigonometric-functions, maths: toppr.com, Retrieved 28 April, 2019
- Triangle-simple-geometry, maths: quatr.us, Retrieved 9 March, 2019
- Triangles-properties-types-geometry: mbacrystalball.com, Retrieved 19 July, 2019
- Circle, geometry, math-vocabulary: splashmath.com, Retrieved 26 January, 2019
- Circle, basic-quadrilateral-ideas, maths: toppr.com, Retrieved 14 May, 2019
- Square, geometry: newworldencyclopedia.org, Retrieved 27 February, 2019
- What-is-a-rectangle, definition, properties, geometry, math: tutors.com, Retrieved 30 July, 2019
- Parallelogram: mathopenref.com, Retrieved 2 August, 2019
- Trapezium-parallelogram-and-rhombus, mensuration, maths: toppr.com, Retrieved 20 March, 2019
- What-is-a-trapezoid, definition, properties, geometry, math: tutors.com, Retrieved 12 April, 2019
- Solid-geometry: onlinemathlearning.com, Retrieved 20 June, 2019

4

Trigonometry

Trigonometry is a mathematical branch that focuses on the study of relationships between the angles and side lengths of a triangle. It is used in various fields such as mechanical engineering, electrical engineering, cartography, computer graphics, etc. This chapter discusses in detail the theories and methodologies related to trigonometry.

Trigonometry is the branch of mathematics concerned with specific functions of angles and their application to calculations. There are six functions of an angle commonly used in trigonometry. Their names and abbreviations are sine (sin), cosine (cos), tangent (tan), cotangent (cot), secant (sec), and cosecant (csc). These six trigonometric functions in relation to a right triangle are displayed in the figure. For example, the triangle contains an angle A, and the ratio of the side opposite to A and the side opposite to the right angle (the hypotenuse) is called the sine of A, or sin A; the other trigonometry functions are defined similarly. These functions are properties of the angle A independent of the size of the triangle, and calculated values were tabulated for many angles before computers made trigonometry tables obsolete. Trigonometric functions are used in obtaining unknown angles and distances from known or measured angles in geometric figures.

Trigonometry developed from a need to compute angles and distances in such fields as astronomy, mapmaking, surveying, and artillery range finding. Problems involving angles and distances in one plane are covered in plane trigonometry. Applications to similar problems in more than one plane of three-dimensional space are considered in spherical trigonometry.

Plane Trigonometry

In many applications of trigonometry the essential problem is the solution of triangles. If enough sides and angles are known, the remaining sides and angles as well as the area can be calculated, and the triangle is then said to be solved. Triangles can be solved by the law of sines and the law of cosines. To secure symmetry in the writing of these laws, the angles of the triangle are lettered A, B, and C and the lengths of the sides opposite the angles are lettered a, b, and c, respectively.

In addition to the angles (A, B, C) and sides (a, b, c), one of the three heights of the triangle (h) is included by drawing the line segment from one of the triangle's vertices (in this case C) that is perpendicular to the opposite side of the triangle.

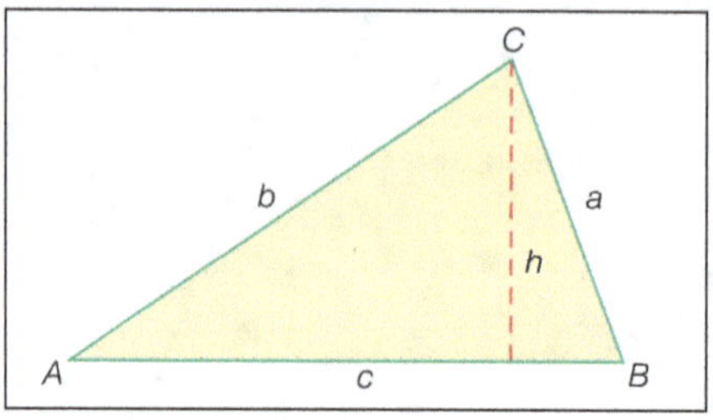

Standard lettering of a triangle

The law of sines is expressed as an equality involving three sine functions while the law of cosines is an identification of the cosine with an algebraic expression formed from the lengths of sides opposite the corresponding angles. To solve a triangle, all the known values are substituted into equations expressing the laws of sines and cosines, and the equations are solved for the unknown quantities. For example, the law of sines is employed when two angles and a side are known or when two sides and an angle opposite one are known. Similarly, the law of cosines is appropriate when two sides and an included angle are known or three sides are known.

Texts on trigonometry derive other formulas for solving triangles and for checking the solution. Older textbooks frequently included formulas especially suited to logarithmic calculation. Newer textbooks, however, frequently include simple computer instructions for use with a symbolic mathematical program.

Spherical Trigonometry

Spherical trigonometry involves the study of spherical triangles, which are formed by the intersection of three great circle arcs on the surface of a sphere. Spherical triangles were subject to intense study from antiquity because of their usefulness in navigation, cartography, and astronomy.

The angles of a spherical triangle are defined by the angle of intersection of the corresponding tangent lines to each vertex. The sum of the angles of a spherical triangle is always greater than the sum of the angles in a planar triangle (π radians, equivalent to two right angles). The amount by which each spherical triangle exceeds two right angles (in radians) is known as its spherical excess. The area of a spherical triangle is given by the product of its spherical excess E and the square of the radius r of the sphere it resides on—in symbols, Er^2.

Common spherical trigonometry formulas

$$\text{law of sines}: \quad \frac{\sin a}{\sin A} = \frac{\sin b}{\sin B} = \frac{\sin c}{\sin C}$$

$$\begin{aligned} &\cos a = \cos b \cos c + \sin b \sin c \cos A \\ \text{law of cosines}: \quad &\cos b = \cos a \cos c + \sin a \sin c \cos B \\ &\cos c = \cos a \cos b + \sin a \sin b \cos C \end{aligned}$$

$$\tan\left(\frac{A}{2}\right)=\sqrt{\frac{\sin(s-b)\sin(s-c)}{\sin s\sin(s-a)}}$$

half-angle formulas: $$\tan\left(\frac{B}{2}\right)=\sqrt{\frac{\sin(s-c)\sin(s-a)}{\sin s\sin(s-b)}}$$

$$\tan\left(\frac{C}{2}\right)=\sqrt{\frac{\sin(s-a)\sin(s-b)}{\sin s\sin(s-c)}}\quad, where\ s=\frac{a+b+c}{2}$$

$$\tan\left(\frac{a}{2}\right)=\sqrt{\frac{-\cos S\cos(S-A)}{\cos(S-B)\cos(S-C)}}$$

half-side formulas: $$\tan\left(\frac{b}{2}\right)=\sqrt{\frac{-\cos S\cos(S-B)}{\cos(S-A)\cos(S-C)}}$$

$$\tan\left(\frac{c}{2}\right)=\sqrt{\frac{-\cos S\cos(S-C)}{\cos(S-A)\cos(S-B)}}\quad, where\ S=\frac{A+B+C}{2}$$

By connecting the vertices of a spherical triangle with the centre O of the sphere that it resides on, a special "angle" known as a trihedral angle is formed. The central angles (also known as dihedral angles) between each pair of line segments OA, OB, and OC are labeled α, β, and γ to correspond to the sides (arcs) of the spherical triangle labeled a, b, and c, respectively. Because a trigonometric function of a central angle and its corresponding arc have the same value, spherical trigonometry formulas are given in terms of the spherical angles A, B, and C and, interchangeably, in terms of the arcs a, b, and c and the dihedral angles α, β, and γ. Furthermore, most formulas from plane trigonometry have an analogous representation in spherical trigonometry. For example, there is a spherical law of sines and a spherical law of cosines.

For a plane triangle, the known values involving a spherical triangle are substituted in the analogous spherical trigonometry formulas, such as the laws of sines and cosines, and the resulting equations are then solved for the unknown quantities.

Napier's analogies
$\tan\left(\frac{a}{2}\right)\cos\left(\frac{B-C}{2}\right)=\tan\left(\frac{b+c}{2}\right)\cos\left(\frac{B+C}{2}\right)$
$\tan\left(\frac{a}{2}\right)\sin\left(\frac{B-C}{2}\right)=\tan\left(\frac{b-c}{2}\right)\sin\left(\frac{B+C}{2}\right)$
$\cot\left(\frac{A}{2}\right)\cos\left(\frac{b-c}{2}\right)=\tan\left(\frac{B+C}{2}\right)\cos\left(\frac{b+c}{2}\right)$
$\cot\left(\frac{A}{2}\right)\sin\left(\frac{b-c}{2}\right)=\tan\left(\frac{B-C}{2}\right)\sin\left(\frac{b+c}{2}\right)$

For each of the included formulas two other formulas can be obtained by permuting the sides and angles: $(a,A)\rightarrow(b,B)\rightarrow(c,C)\rightarrow(a,A)$.

Many other relations exist between the sides and angles of a spherical triangle. Worth mentioning are Napier's analogies (derivable from the spherical trigonometry half-angle or half-side formulas), which are particularly well suited for use with logarithmic tables.

Analytic Trigonometry

Analytic trigonometry combines the use of a coordinate system, such as the Cartesian coordinate system used in analytic geometry, with algebraic manipulation of the various trigonometry functions to obtain formulas useful for scientific and engineering applications.

Trigonometric functions of a real variable x are defined by means of the trigonometric functions of an angle. For example, $\sin x$ in which x is a real number is defined to have the value of the sine of the angle containing x radians. Similar definitions are made for the other five trigonometric functions of the real variable x. These functions satisfy the trigonometric relations with A, B, 90°, and 360° replaced by x, y, $\pi/2$ radians, and 2π radians, respectively. The minimum period of $\tan x$ and $\cot x$ is π, and of the other four functions it is 2π.

In calculus it is shown that $\sin x$ and $\cos x$ are sums of power series. These series may be used to compute the sine and cosine of any angle. For example, to compute the sine of 10°, it is necessary to find the value of $\sin \pi/18$ because 10° is the angle containing $\pi/18$ radians. When $\pi/18$ is substituted in the series for $\sin x$, it is found that the first two terms give 0.17365, which is correct to five decimal places for the sine of 10°. By taking enough terms of the series, any number of decimal places can be correctly obtained. Tables of the functions may be used to sketch the graphs of the functions.

Each trigonometric function has an inverse function, that is, a function that "undoes" the original function. For example, the inverse function for the sine function is written arcsin or $\sin^{-1}$, thus $\sin^{-1}(\sin x) = \sin(\sin^{-1} x) = x$. The other trigonometric inverse functions are defined similarly.

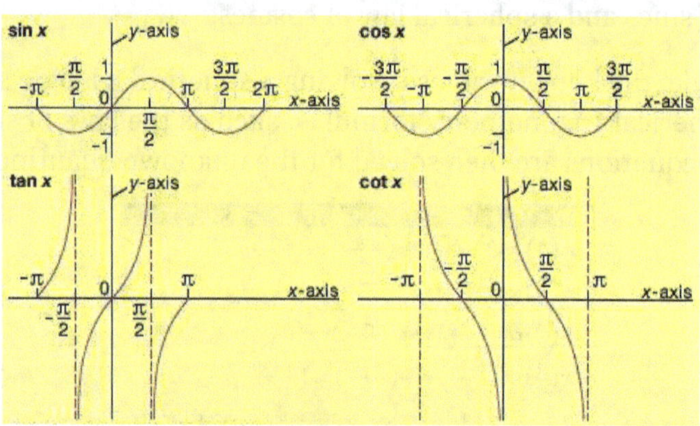

Graphs of some trigonometric functions. Note that each of these functions is periodic. Thus, the sine and cosine functions repeat every 2π, and the tangent and cotangent functions repeat every π.

Coordinates and Transformation of Coordinates

Polar Coordinates

For problems involving directions from a fixed origin (or pole) O, it is often convenient to specify a point P by its polar coordinates (r, θ), in which r is the distance OP and θ is the angle that the direction of r makes with a given initial line. The initial line may be identified with the x-axis of rectangular Cartesian coordinates, as shown in the figure. The point (r, θ) is the same as $(r, \theta + 2n\pi)$ for any integer n. It is sometimes desirable to allow r to be negative, so that (r, θ) is the same as $(-r, \theta + \pi)$.

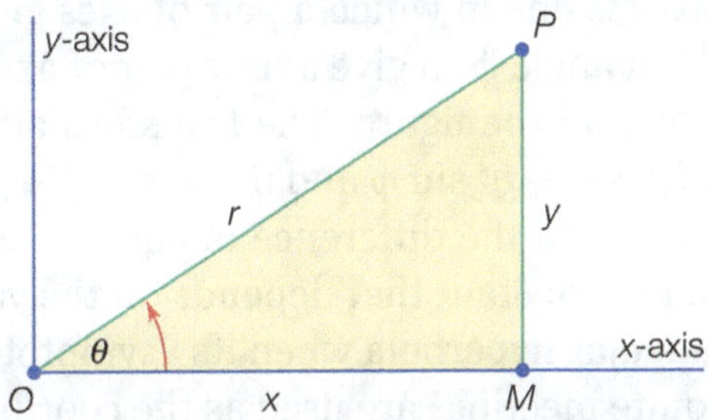

Cartesian and polar coordinates

The point labeled P in the figure resides in the plane. Therefore, it requires two dimensions to fix its location, either in Cartesian coordinates (x, y) or in polar coordinates (r, θ).

Given the Cartesian equation for a curve, the polar equation for the same curve can be obtained in terms of the radius r and the angle θ by substituting $r \cos \theta$ and $r \sin \theta$ for x and y, respectively. For example, the circle $x^2 + y^2 = a^2$ has the polar equation $(r \cos \theta)^2 + (r \sin \theta)^2 = a^2$, which reduces to $r = a$. (The positive value of r is sufficient, if θ takes all values from $-\pi$ to π or from 0 to 2π). Thus the polar equation of a circle simply expresses the fact that the curve is independent of θ and has constant radius. In a similar manner, the line $y = x \tan \phi$ has the polar equation $\sin \theta = \cos \theta \tan \phi$, which reduces to $\theta = \phi$. (The other solution, $\theta = \phi + \pi$, can be discarded if r is allowed to take negative values.)

Transformation of Coordinates

A transformation of coordinates in a plane is a change from one coordinate system to another. Thus, a point in the plane will have two sets of coordinates giving its position with respect to the two coordinate systems used, and a transformation will express the relationship between the coordinate systems. For example, the transformation between polar and Cartesian coordinates is given by $x = r \cos \theta$ and $y = r \sin \theta$. Similarly, it is possible to accomplish transformations between rectangular and oblique coordinates.

In a translation of Cartesian coordinate axes, a transformation is made between two sets of axes that are parallel to each other but have their origins at different positions.

If a point P has coordinates (x, y) in one system, its coordinates in the second system are given by $(x - h, y - k)$ where (h, k) is the origin of the second system in terms of the first coordinate system. Thus, the transformation of P between the first system (x, y) and the second system (x', y') is given by the equations $x = x' + h$ and $y = y' + k$. The common use of translations of axes is to simplify the equations of curves. For example, the equation $2x^2 + y^2 - 12x - 2y + 17 = 0$ can be simplified with the translations $x' = x - 3$ and $y' = y - 1$ to an equation involving only squares of the variables and a constant term: $(x')^2 + (y')^2/2 = 1$. In other words, the curve represents an ellipse with its centre at the point (3, 1) in the original coordinate system.

A rotation of coordinate axes is one in which a pair of axes giving the coordinates of a point (x, y) rotate through an angle ϕ to give a new pair of axes in which the point has coordinates (x', y'), as shown in the figure. The transformation equations for such a rotation are given by $x = x' \cos \phi - y' \sin \phi$ and $y = x' \sin \phi + y' \cos \phi$. The application of these formulas with $\phi = 45°$ to the difference of squares, $x^2 - y^2 = a^2$, leads to the equation $x'y' = c$ (where c is a constant that depends on the value of a). This equation gives the form of the rectangular hyperbola when its asymptotes (the lines that a curve approaches without ever quite meeting) are used as the coordinate axes.

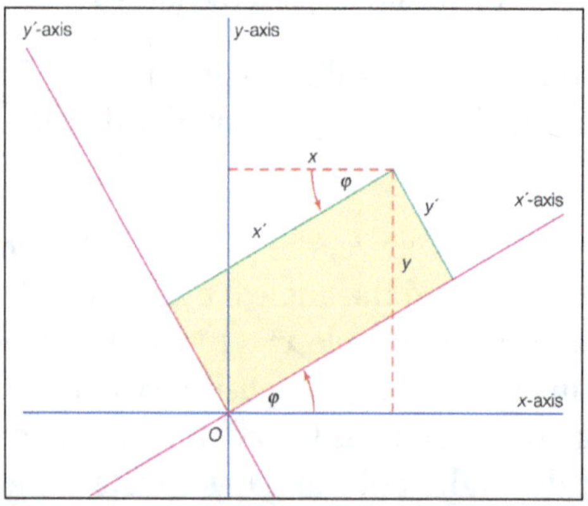

Rotation of axes. Rotating the coordinate axes through an angle ϕ changes the coordinates of a point from (x, y) to (x', y').

Trigonometric Functions

Trigonometric functions are also known as a Circular Functions can be simply defined as the functions of an angle of a triangle i.e. the relationship between the angles and sides of a triangle are given by these trig functions. The formulas, table and definition of basic functions such as sin, cos and tan are given here. Also, the other three ratios like sec, cosec and cot, which can be represented in graphs as well, have been explained here. There are a number of trigonometric formula and identities which denotes the relation between the functions and help to find the angles of the triangle.

Also, you will come across the table for where the value of these ratios is mentioned for some particular degrees. And based on this table you will be able to solve many trigonometric examples and problems.

Sin, Cos and Tan Functions

The angles of sine, cosine, and tangent are the primary classification of functions of trigonometry. And the three functions which are cotangent, secant and cosecant can be derived from the primary functions. Basically, the other three functions are often used as compare to the primary trigonometric functions. Consider the following diagram as a reference for an explanation of these three primary functions. This diagram can be referred to as the sin-cos-tan triangle. We usually define the trigonometry with the help of the right-angled triangle.

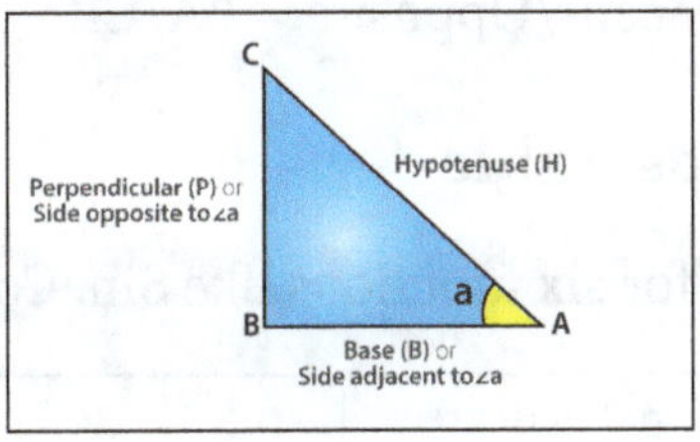

Trigonometry Functions Formula

The formulas for functions of trigonometric ratios(sine, cosine and tangent) for a right-angled triangle:

Sine function:

Sine function of an angle is the ratio between the opposite side length to that of the hypotenuse. From the above diagram, the value of sin will be:

Sin a =Opposite/Hypotenuse = CB/CA

Cos function:

Cos of an angle is the ratio of the length of the adjacent side to the length of the hypotenuse. From the above diagram, the cos function will be derived as follows.

Cos a = Adjacent/Hypotenuse = AB/CA

Tan function:

The tangent function is the ratio of the length of the opposite side to that of the adjacent side. It should be noted that the tan can also be represented in terms of sine and cos as their ratio. From the diagram taken above, the tan function will be the following.

Tan a = Opposite/Adjacent = CB/BA

Also, in terms of sine and cos, tan can be represented as:

Tan a = sin a/cos a

Secant, cosecant and cotangent functions:

Secant, cosecant (csc) and cotangent are the three additional functions which are derived from the primary functions of sine, cos, and tan. The reciprocal of sine, cos, and tan are cosecant (csc), secant (sec), and cotangent (cot) respectively. The formula of each of these functions are given as:

Sec a = 1/(cos a) = Hypotenuse/Adjacent = CA/AB

Cosec a = 1/(sin a) = Hypotenuse/Opposite = CA/CB

cot a = 1/(tan a) = Adjacent/Opposite = BA/CB

Trigonometric Functions Table

The trigonometric ratio table for six functions like Sin, Cos, Tan, Cosec, Sec, Cot, are:

Trigonometric Ratios/angle= θ in degrees	0 °	30 °	45 °	60 °	90 °
Sin θ	0	1/2	1/√2	√3/2	1
Cos θ	1	√3/2	1/√2	1/2	0
Tan θ	0	1/√3	1	√3	∞
Cosec θ	∞	2	√2	2/√3	1
Sec θ	1	2/√3	√2	2	∞
Cot θ	∞	√3	1	1/√3	0

Inverse functions are used to obtain an angle from any of the angle's trigonometric ratios. Basically, inverses of the sine, cosine, tangent, cotangent, secant, and cosecant functions are represented as arcsine, arccosine, arctangent, arccotangent, arcsecant, and arc cosecant.

Trigonometric Function Examples

Example: Find the values of Sin 45°, Cos 30° and Tan 60°.[all angles are in degrees)

Solution: Using the trigonometric table, we have

Sin 45° = 1/√2

Cos 30° = 1/2

Tan 60° = √3

Example: Evaluate Sin 105° degrees.

Solution: Sin 105° can be written as sin (60° + 45°) which is similar to sin (A + B).

We know that, the formula for sin (A + B) = sin A × cos B + cos A × sin B

Therefore, sin 105° = sin (60° + 45°) = sin 60° × cos 45° + cos 60° × sin 45°

= √3/2 × 1/√2 + 1/2 × 1/√2

= √3/2√2 + 1/2√2

= (√2 + √6)/4

Example: A boy sees a bird sitting on a tree at an angle of elevation of 20°. If a boy is standing 10 miles away from the tree, at what height bird is sitting?

Solution: Consider ABC a right triangle, A is a bird's location, B = tree is touching the ground and C = boy's location.

So BC 10 miles, angle C = 20° and let AB = x miles

We know, tan C = opposite side/adjacent side

tan(20°) = x/10

or x = 10 × tan(20°)

or x = 10 × 0.36 = 3.6

Bird is sitting at the height of 3.6 miles from the ground.

Inverse Trigonometric Functions

The inverse trigonometric functions are the inverse functions of the trigonometric functions, written $\cos^{-1} z, \cot^{-1} z, \csc^{-1} z, \sec^{-1} z, \sin^{-1} z$, and $\tan^{-1} z$.

Alternate notations are sometimes used, as summarized in the following table.

$f(z)$	Alternate Notations
$\cos^{-1} z$	arccos z
$\cot^{-1} z$	arccot z
$\csc^{-1} z$	arccs z
$\sec^{-1} z$	arcsec z

$\sin^{-1} z$	arcsin z
$\tan^{-1} z$	arctan z

The inverse trigonometric functions are multivalued. For example, there are multiple values of w such that $z = \sin w$, so $\sin^{-1} z$ is not uniquely defined unless a principal value is defined. Such principal values are sometimes denoted with a capital letter so, for example, the principal value of the inverse sine $\sin^{-1} z$ may be variously denoted $\sin^{-1} z$ or Arcsin z. On the other hand, the notation $\sin^{-1} z$ (etc.) is also commonly used denote either the principal value or *any* quantity whose sine is z an. Worse still, the principal value and multiply valued notations are some-times reversed, with for example arcsin z denoting the principal value and Arcsin z denoting the multivalued functions.

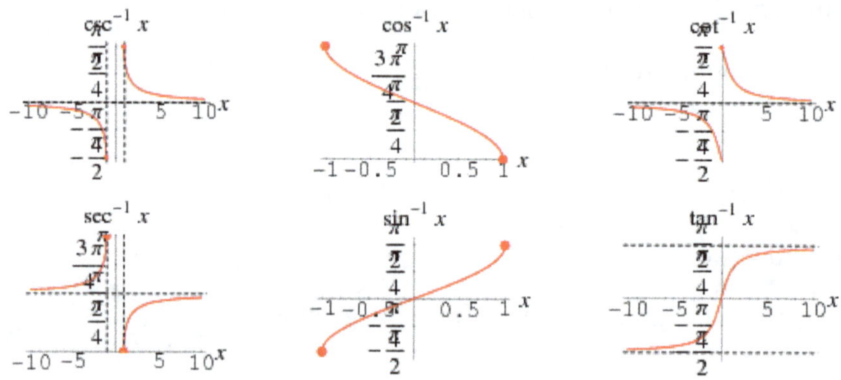

Different conventions are possible for the range of these functions for real arguments. Following the convention used, the inverse trigonometric functions defined in this work have the following ranges for domains on the real line ℝ, illustrated above.

Function Name	Function	Domain	Range
inverse cosecant	$\csc^{-1} x$	$(-\infty, \infty)$	$\left[-\frac{1}{2}\pi, 0) \text{ or } (0, \frac{1}{2}\pi\right]$
inverse cosine	$\cos^{-1} x$	$[-1, 1]$	$[0, \pi]$
inverse cotangent	$\cot^{-1} x$	$(-\infty, \infty)$	$\left[-\frac{1}{2}\pi, 0) \text{ or } (0, \frac{1}{2}\pi\right]$
inverse secant	$\sec^{-1} x$	$(-\infty, \infty)$	$\left[-\frac{1}{2}\pi,) \text{ or } (\frac{1}{2}\pi, \pi\right]$

inverse sine	$\sin^{-1} x$	$[-1,1]$	$\left[-\frac{1}{2}\pi, \frac{1}{2}\pi\right]$
inverse tangent	$\tan^{-1} x$	$(-\infty,\infty)$	$\left[-\frac{1}{2}\pi, \frac{1}{2}\pi\right]$

Inverse-forward identities are

$$\tan^{-1}(\cot x) = \frac{1}{2}\pi - x \text{ for } x \in [0,\pi]$$

$$\sin^{-1}(\cot x) = \frac{1}{2}\pi - x \text{ for } x \in [0,\pi]$$

$$\sec^{-1}(\csc x) = \frac{1}{2}\pi - x \text{ for } x \in \left[0, \frac{1}{2}\pi\right].$$

Forward-inverse identities are

$$\cos\left(\sin^{-1} x\right) = \sqrt{1-x^2}$$

$$\cos\left(\tan^{-1} x\right) = \frac{1}{\sqrt{1+x^2}}$$

$$\sin\left(\cos^{-1} x\right) = \sqrt{1-x^2}$$

$$\sin\left(\tan^{-1} x\right) = \frac{x}{\sqrt{1+x^2}}$$

$$\tan\left(\cos^{-1} x\right) = \frac{\sqrt{1-x^2}}{x}$$

$$\tan\left(\sin^{-1} x\right) = \frac{x}{\sqrt{1-x^2}}.$$

Inverse sum identities include

$$\sin^{-1} x + \cos^{-1} x = \frac{1}{2}\pi$$

$$\tan^{-1} x + \cot^{-1} x = \frac{1}{2}\pi$$

$$\sec^{-1} x + \csc^{-1} x = \frac{1}{2}\pi,$$

where equation $\tan^{-1} x + \cot^{-1} x = \frac{1}{2}\pi$ is valid only for $x \geq 0$.

Complex inverse identities in terms of natural logarithms include

$$\sin^{-1} z = -i\,\text{In}\left(iz+\sqrt{1-z^2}\right)$$

$$\cos^{-1} z = \frac{1}{2}\pi + i\,\text{In}\left(iz+\sqrt{1-z^2}\right)$$

$$\tan^{-1} z = \frac{1}{2}i\left[\text{In}(1-iz)-\text{In}(1+iz)\right].$$

Properties of Trigonometric Inverse Functions

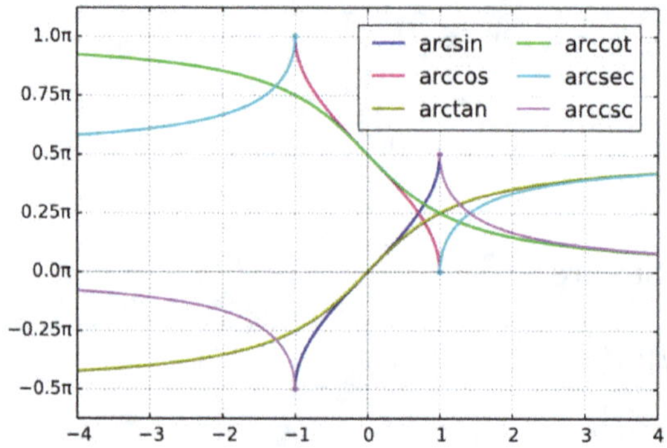

Property 1

i. $\sin^{-1}(1/x) = \text{cosec}^{-1}x$, $x \geq 1$ or $x \leq -1$

ii. $\cos^{-1}(1/x) = \sec^{-1}x$, $x \geq 1$ or $x \leq -1$

iii. $\tan^{-1}(1/x) = \cot^{-1}x$, $x > 0$

Proof : $\sin^{-1}(1/x) = \text{cosec}^{-1}x$, $x \geq 1$ or $x \leq -1$,

Let sin−1x=y

i.e. $x = \text{cosec } y$

$$\frac{1}{x} = \sin y$$

$$\sin^{-1}\left(\frac{1}{x}\right) = y$$

$$\sin^{-1}\left(\frac{1}{x}\right) = \text{cosec}^{-1} x$$

$$\sin^{-1}\left(\frac{1}{x}\right) = \text{cosec}^{-1} x$$

Hence, $\sin^{-1}\frac{1}{x} = \text{cosec}^{-1} x$ where, $x \geq 1$ or $x \leq -1$.

Property 2

i. $\sin^{-1}(-x) = -\sin^{-1}(x), x \in [-1,1]$

ii. $\tan^{-1}(-x) = -\tan^{-1}(x), x \in R$

iii. $\text{cosec}^{-1}(-x) = -\text{cosec}^{-1}(x), |x| \geq 1$

Proof: $\sin^{-1}(-x) = -\sin^{-1}(x), x \in [-1,1]$

Let, $\sin^{-1}(-x)=y$

Then $-x=\sin y$

$x=-\sin y$

$x=\sin(-y)$

$\sin^{-1}=\sin^{-1}(\sin(-y))$

$\sin^{-1} x=y$

$\sin^{-1} x=-\sin^{-1}(-x)$

Hence, $\sin^{-1}(-x)=-\sin^{-1} x \in [-1,1]$

Property 3

i. $\cos^{-1}(-x) = \pi - \cos^{-1} x, x \in [-1,1]$

ii. $\sec^{-1}(-x) = \pi - \sec^{-1}x, |x| \geq 1$

iii. $\cot^{-1}(-x) = \pi - \cot^{-1}x, x \in R$

Proof : $\cos^{-1}(-x) = \pi - \cos^{-1} x, x \in [-1,1]$

Let $\cos^{-1}(-x)=y$

$\cos y=-x$

$x=-\cos y$

$x=\cos(\pi-y)$

Since, $\cos\pi-q=-\cos q$

$\cos^{-1} x=\pi-y$

$\cos^{-1} x=\pi-\cos^{-1}-x$

Hence, $\cos^{-1}-x=\pi-\cos^{-1}x$

Property 4

i. $\sin^{-1}x + \cos^{-1}x = \pi/2, x \in [-1,1]$

ii. $\tan^{-1}x + \cot^{-1}x = \pi/2, x \in R$

iii. $\operatorname{cosec}^{-1}x + \sec^{-1}x = \pi/2$, $|x| \geq 1$

Proof : $\sin^{-1}x + \cos^{-1}x = \pi/2$, $x \in [-1,1]$

$$\text{Let} \sin^{-1} x = y \,\text{or}\, x = \sin y = \cos\left(\frac{\pi}{2} - y\right)$$

$$\cos^{-1} x = \cos^{-1}\left(\cos\left(\frac{\pi}{2} - y\right)\right)$$

$$\cos^{-1} x = \frac{\pi}{2} - \sin^{-1} x$$

$$\sin^{-1} + \cos^{-1} x = \frac{\pi}{2}$$

$$\text{Hence}, \sin^{-1} \text{x} + \cos^{-1} \text{x} = \pi / 2, \text{x} \in [-1,1]$$

Property 5

i. $\tan^{-1}x + \tan^{-1}y = \tan^{-1}((x+y)/(1-xy))$, $xy < 1$.

ii. $\tan^{-1}x - \tan^{-1}y = \tan^{-1}((x-y)/(1+xy))$, $xy > -1$.

Proof : $\tan^{-1}x + \tan^{-1}y = \tan^{-1}((x+y)/(1-xy))$, $xy < 1$.

Let $\tan^{-1}x=A$

And $\tan^{-1}y=B$

Then, $\tan A=x$

$\tan B=y$

Now, $\tan(A+B)=(\tan A+\tan B)/(1-\tan A \tan B)$

$$\tan(A+B) = \frac{x+y}{1-xy}$$

$$\tan^{-1}\left(\frac{x+y}{1-xy}\right) = A+B$$

$$\text{Hence}, \tan^{-1}\left(\frac{x+y}{1-xy}\right) = \tan^{-1} x + \tan^{-1} y$$

Property 6

i. $2\tan^{-1}x = \sin^{-1}(2x/(1+x^2))$, $|x| \leq 1$

ii. $2\tan^{-1}x = \cos^{-1}((1-x^2)/(1+x^2)),\ x \geq 0$

iii. $2\tan^{-1}x = \tan^{-1}(2x/(1-x^2)),\ -1 < x < 1$

Proof : $2\tan^{-1}x = \sin^{-1}(2x/(1+x^2)),\ |x| \leq 1$

Let $\tan^{-1}x=y$ and $x=\tan y$

$$\text{Consider RHS}.\sin^{-1}\left(\frac{2x}{1+x^2}\right)$$
$$=\sin^{-1}\left(\frac{2\tan y}{1+\tan^2 y}\right)$$
$$=\sin^{-1}(\sin 2y)$$
$$\text{Since}, \sin 2\theta = 2\tan\theta/1\left(1+\tan^2\theta\right),$$

$$=2y$$
$$=2\tan^{-1}x \text{ which is our LHS}$$
$$\textit{Hence}\ 2\tan^{-1}x = \sin^{-1}\left(2x/\left(1+x^2\right)\right), |x| \leq 1$$

Example: Prove that "$\sin^{-1}(-x) = -\sin^{-1}(x),\ x \in [-1,1]$"

Solution: Let, $\sin^{-1}(-x)=y$

Then $-x=\sin y$

$x=-\sin y$

$x=\sin(-y)$

$\sin^{-1}x=\arcsin(\sin(-y))$

$\sin^{-1}x=y$

$\sin^{-1}x=-\sin^{-1}(-x)$

Hence, $\sin^{-1}(-x)=-\sin^{-1}x,\ x \in [-1,1]$

Derivatives of Trigonometric Functions

Our starting point is the following limit:

$$\lim_{x\to 0}\frac{\sin(x)}{x}=1.$$

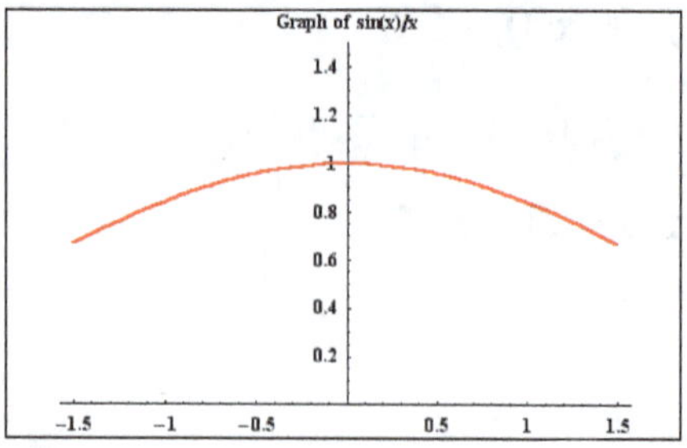

Using the derivative language, this limit means that $\sin'(0)=1$. This limit may also be used to give a related one which is of equal importance:

$$\lim_{x\to 0}\frac{\cos(x)-1}{x}=0.$$

To see why, it is enough to rewrite the expression involving the cosine as

$$\frac{\cos(x)-1}{x}=\frac{(\cos(x)-1)(\cos(x)+1)}{x(\cos(x)+1)}=\frac{(\cos^2(x)-1)}{x(\cos(x)+1)}$$

But $\cos^2(x)-1=-\sin^2(x)$, so we have

$$\lim_{x\to 0}\frac{\cos(x)-1}{x}=\lim_{x\to 0}\frac{-\sin^2(x)}{x(\cos(x)+1)}=\lim_{x\to 0}x\frac{-\sin^2(x)}{x^2(\cos(x)+1)}=0.$$

This limit equals $\cos'(0)$ and thus $\cos'(0)=0$.

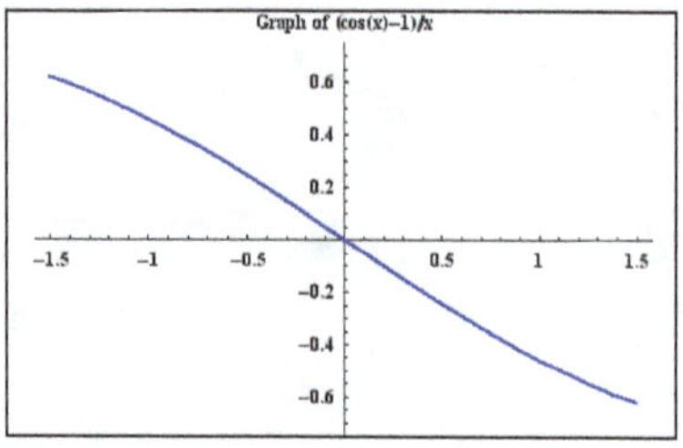

In fact, we may use these limits to find the derivative of $\sin(x)$ and $\cos(x)$ at any point $x=a$. Indeed, using the addition formula for the sine function, we have

$$\sin(a+h)=\sin(a)\cos(h)+\sin(h)\cos(a).$$

So

$$\frac{\sin(a+h)-\sin(a)}{h}=\sin(a)\frac{\cos(h)-1}{h}+\cos(a)\frac{\sin(h)}{h}$$

which implies

$$\lim_{x \to 0} \frac{\sin(a+h)-\sin(a)}{h} = \cos(a).$$

So we have proved that $\sin'(a)$ exists and $\sin'(a)=\cos(a)$.

Similarly, we obtain that $\cos'(a)$ exists and that $\cos'(a)=-\sin(a)$.

Since $\tan(x)$, $\cot(x)$, $\sec(x)$, and $\csc(x)$ are all quotients of the functions $\sin(x)$ and $\cos(x)$, we can compute their derivatives with the help of the quotient rule:

$$\frac{d}{dx}(\tan(x)) = \sec^2(x) = 1+\tan^2(x) \quad \frac{d}{dx}(\sec(x)) = \sec(x)\tan(x)$$

$$\frac{d}{dx}(\cot(x)) = -\csc^2(x) = -1-\cot^2(x) \quad \frac{d}{dx}(\csc(x)) = -\csc(x)\cot(x)$$

It is quite interesting to see the close relationship between $\tan(x)$ and $\sec(x)$ (and also between $\cot(x)$ and $\csc(x)$).

From the above results we get

$$\sin''(x)-\sin(x) \text{ and } \cos''(x)=-\cos(x).$$

These two results are very useful in solving some differential equations.

Example: Let $f(x)=\sin(2x)$. Using the double angle formula for the sine function, we can rewrite

$$\sin(2x)=2\sin(x)\cos(x).$$

So using the product rule, we get

$$\frac{d}{dx}(\sin(2x)) = 2(\cos(x)\cos(x)-\sin(x)\sin(x)) = 2(\cos^2(x)-\sin^2(x))$$

which implies, using trigonometric identities,

$$\frac{d}{dx}(\sin(2x)) = 2\cos(2x).$$

Laws of Sines

The Law of Sines is the relationship between the sides and angles of non-right (oblique) triangles. Simply, it states that the ratio of the length of a side of a triangle to the sine of the angle opposite that side is the same for all sides and angles in a given triangle.

In ΔABC is an oblique triangle with sides a,b and c, then $\frac{a}{\sin A}=\frac{b}{\sin B}=\frac{c}{\sin C}$.

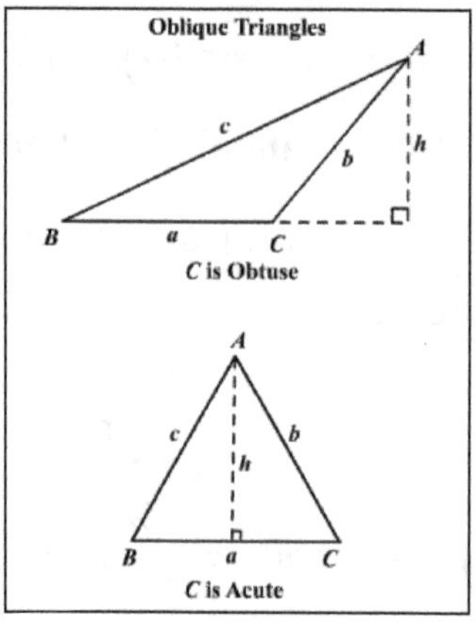

To use the Law of Sines you need to know either two angles and one side of the triangle (AAS or ASA) or two sides and an angle opposite one of them (SSA). Notice that for the first two cases we use the same parts that we used to prove congruence of triangles in geometry but in the last case we could not prove congruent triangles given these parts. This is because the remaining pieces could have been different sizes. This is called the ambiguous case.

Example: Given two angles and a non-included side (AAS).

Given ΔABC with $m\angle A=30°$, $m\angle B=20°$ *and* $a=45$ m. Find the remaining angle and sides.

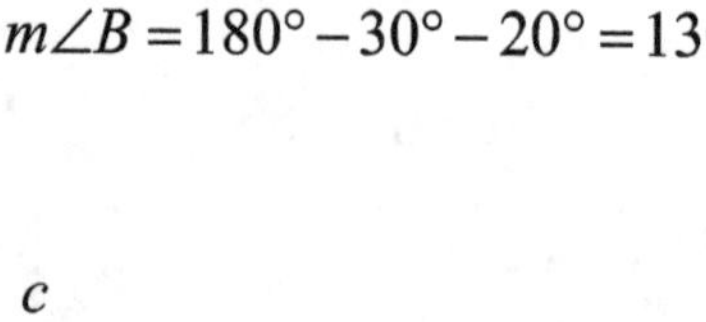

The third angle of the triangle is

$$m\angle C=180°-m\angle A-m\angle B=180°-30°-20°=130°$$

By the law of sines,

$$\frac{45}{\sin 30°}=\frac{b}{\sin 20°}=\frac{c}{\sin 130°}$$

By the properties of proportions

$$b=\frac{45\sin 20°}{\sin 30°}\approx 30.78 \text{ m and } c=\frac{45\sin 130°}{\sin 30°}\approx 68.94 \text{ m}$$

Example: Given two angles and an included side (ASA).

Given $m\angle A = 42°$ $m\angle B = 75°$ and $c = 22$ cm. Find the remaining angle and sides.

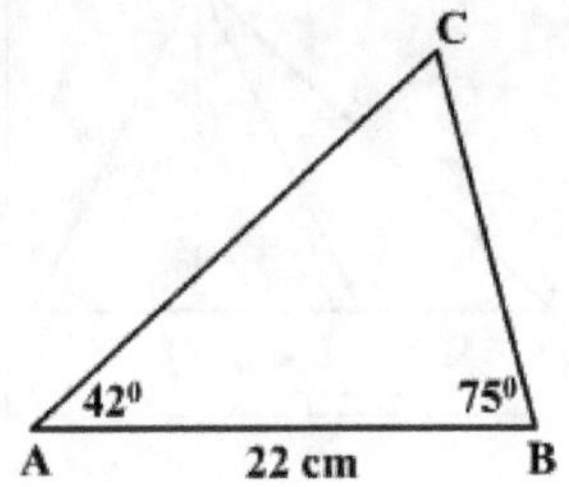

The third angle of the triangle is:

$$m\angle C = 180° - m\angle A - m\angle B = 180° - 42° - 75° = 63°$$

By the law of sines,

$$\frac{a}{\sin 42°} = \frac{b}{\sin 75°} = \frac{22}{\sin 63°}$$

By the properties of proportions

$$a = \frac{22\sin 42°}{\sin 63°} \approx 16.52 \text{ cm and } b = \frac{22\sin 75°}{\sin 63°} \approx 23.85\,cm$$

The Ambiguous Case

If two sides and an angle opposite one of them are given, three possibilities can occur.

(1) No such triangle exists.

(2) Two different triangles exist.

(3) Exactly one triangle exists.

Consider a triangle in which you are given a, b and A. (The altitude h from vertex B to side $\overline{AC}$, by the definition of sines is equal to $bsinA$.)

(1) No such triangle exists if A is acute and $a < h$ or A is obtuse and $a \le b$.

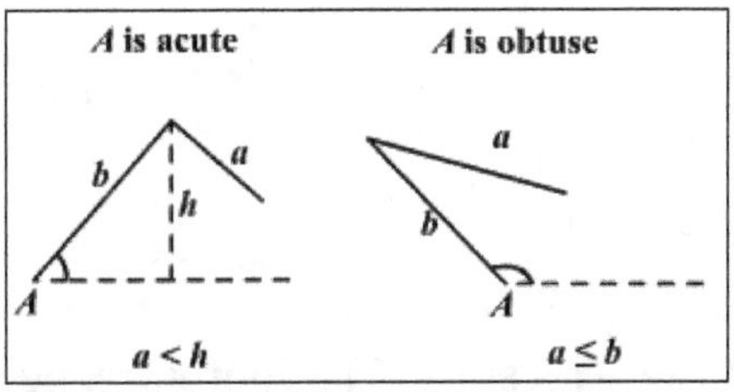

(2) Two different triangles exist if A is acute and $h<a<b$.

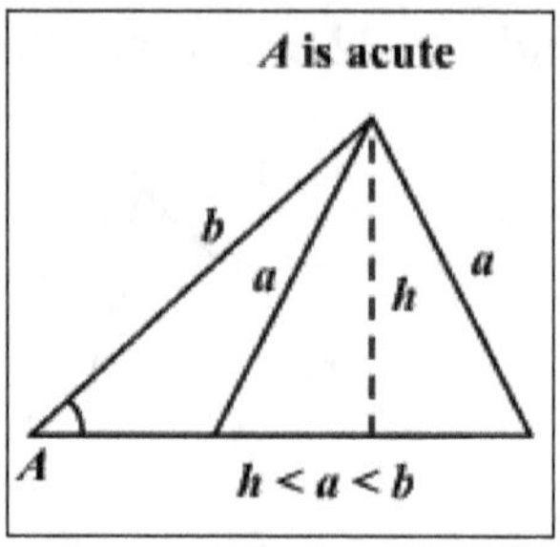

(3) In every other case, exactly one triangle exists.

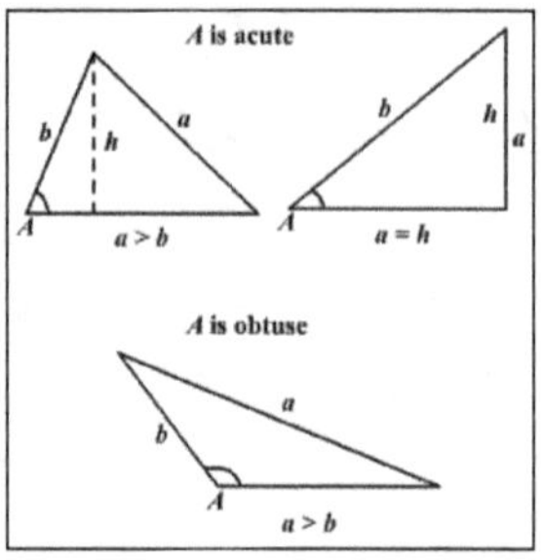

Example: No Solution Exists

Given $a=15, b=25$ and $m\angle A=80°$. Find the other angles and side.

$$h=b\sin A=25\sin 80°\approx 24.6$$

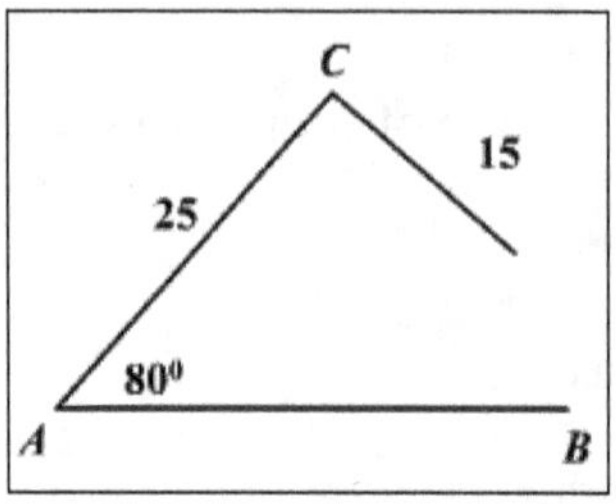

Notice that $a<h$. So it appears that there is no solution. Verify this using the Law of Sines.

$$\frac{a}{\sin A}=\frac{b}{\sin B}$$

$$\frac{15}{\sin 80°}=\frac{25}{\text{sinB}}$$

$$\sin B=\frac{25\sin 80°}{15}\approx 1.641>1$$

This contrasts the fact that the $-1\le \sin B\le 1$. Therefore, no triangle exists.

Example: Two Solutions Exist

Given $a=6, b=7$ and $m\angle A=30°$. Find the other angles and side.

$$h=b\sin A=7\sin 30°=3.5$$

$h<a<b$ therefore, there are two triangles possible.

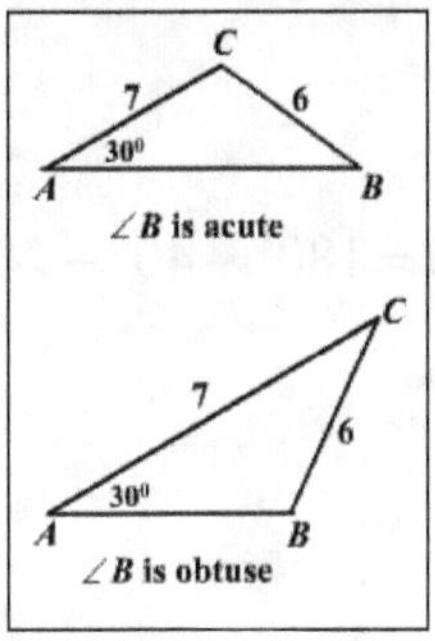

By the Law of Sines, $\dfrac{a}{sinA}=\dfrac{b}{sinB}$

$$\sin B=\frac{b\sin A}{a}=\frac{7\sin 30°}{6}\approx 0.5833$$

There are two angles between $0°$ and $180°$ whose sine is approximately 0.5833, are $35.69°$ and $144.31°$.

If $B\approx 35.69°$

$C\approx 180°-30°-35.69°=114.31°$

$$c=\frac{a\sin C}{\sin}\approx\frac{6\sin 114.31°}{\sin 30°}\approx 10.94$$

If $B\approx 144.31°$

$C\approx 180°-30°-144.31°=5.69°$

$$c\approx\frac{6\sin 5.69°}{\sin 30°}\approx 1.19$$

Example: One Solution Exists

Given $a=22, b=12$ and $m\angle A=40°$. Find the other angles and side.

$a>b$

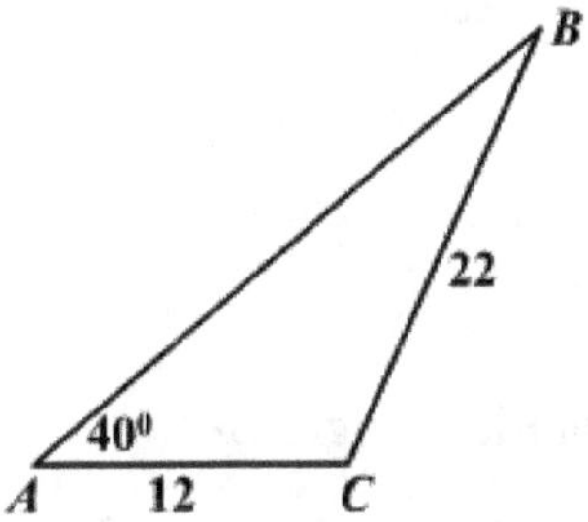

By the Law of Sines, $\dfrac{a}{sinA} = \dfrac{b}{sinB}$

$$\sin B = \frac{b\sin A}{a} = \frac{12\sin 40^\circ}{22} \approx 0.3506$$
$$B \approx 20.52^\circ$$

B is acute.

$$m\angle C = 180^\circ - m\angle A - m\angle B = 180^\circ - 40^\circ - 20.52^\circ = 29.79^\circ$$

By the Law of Sines,

$$\frac{c}{\sin 119.48^\circ} = \frac{22}{\sin 40^\circ}$$
$$c = \frac{22\sin 119.48^\circ}{\sin 40^\circ} \approx 29.79$$

Law of Cosines

The Law of Cosines is used to find the remaining parts of an oblique (non-right) triangle when either the lengths of two sides and the measure of the included angle is known (SAS) or the lengths of the three sides (SSS) are known. In either of these cases, it is impossible to use the Law of Sines because we cannot set up a solvable proportion.

The Law of Cosines states:

$$c^2 = a^2 + b^2 - 2ab\cos C.$$

This resembles the Pythagorean Theorem except for the third term and if C is a right angle the third term equals 0 because the cosine of 90° is 0 and we get the Pythagorean Theorem. So, the Pythagorean Theorem is a special case of the Law of Cosines.

The Law of Cosines can also be stated as

$$b^2 = a^2 + c^2 - 2ac\cos B \text{ or}$$

$$a^2 = b^2 + c^2 - 2bc\cos A$$

Example: Two Sides and the Included Angle-SAS

Given $a = 11, b = 5$ and $m\angle C = 20^\circ$. Find the remaining side and angles.

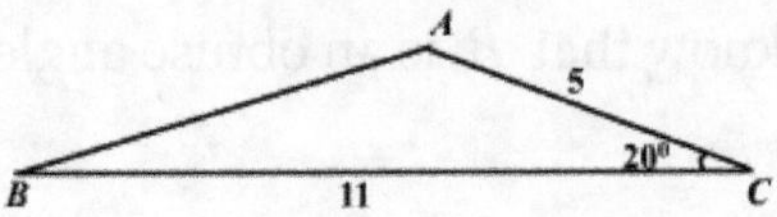

$$c^2 = a^2 + b^2 - 2ab\cos C$$

$$c = \sqrt{a^2 + b^2 - 2ab\cos C}$$

$$= \sqrt{11^2 + 5^2 - 2(11)(5)(cos 20°)}$$

$$\approx 6.53$$

To find the remaining angles, it is easiest to now use the Law of Sines.

$$\sin A \approx \frac{11\sin 20°}{6.53}$$

$$A \approx 144.82°$$

$$\sin B \approx \frac{5\sin 20°}{6.53},$$

$$B \approx 15.2°$$

Note that angle A is opposite to the longest side and the triangle is not a right triangle. So, when you take the inverse you need to consider the obtuse angle whose sine is $\frac{11 sin(20°)}{6.53} \approx 0.5761$.

Example : Three Sides-SSS

Given $a = 8, b = 19$ and $c = 14$. Find the measures of the angles.

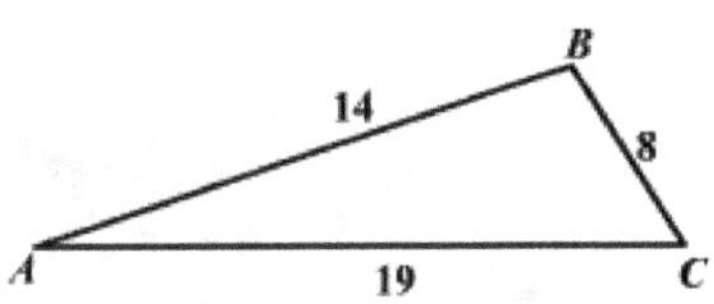

It is best to find the angle opposite the longest side first. In this case, that is side b.

$$\cos B = \frac{b^2 - a^2 - c^2}{-2ac} = \frac{19^2 - 8^2 - 14^2}{-2(8)(14)} \approx -0.45089$$

Since $\cos B$ is negative, we know that B is an obtuse angle.

$$B \approx 116.80°$$

Since B is an obtuse angle and a triangle has at most one obtuse angle, we know that angle A and angle C are both acute.

To find the other two angles, it is simplest to use the Law of Sines.

$$\frac{a}{\sin A} = \frac{b}{\sin B} = \frac{c}{\sin C}$$

$$\frac{8}{\sin A} \approx \frac{19}{\sin 116.80°} \approx 14 sinC$$

$$\sin A \approx \frac{8 \sin 116.80°}{19}$$

$$A \approx 22.08°$$

$$\sin C \approx \frac{14 \sin 116.80°}{19}$$

$$C \approx 41.12°$$

Real Life Applications of Trigonometry

Trigonometry to Measure Height of a Building or a Mountain

Trigonometry is used to in measuring the height of a building or a mountain. The distance of a building from the viewpoint and the elevation angle can easily determine the height of a building using the trigonometric functions.

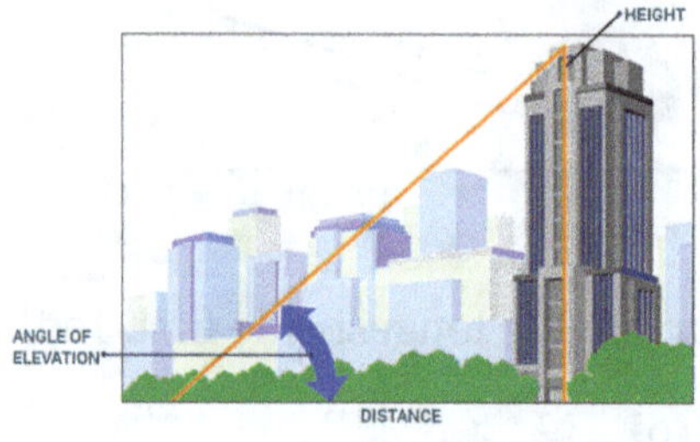

Example: The distance from where the building is observed is 90ft from its base and the angle of elevation to the top of the building is $35°$. Now find the height of the building.

Solution:

Given:

1. Distance from the building is 90feet from the building.
2. The angle of elevation from to the top of the building is $35°$.

To solve and find the height of the tower by recalling the trigonometric formulas. Here, the angle and the adjacent side length are provided. So, using the formula of tan.

$$\tan 35° = \frac{OppositeSide}{AdjacentSide}$$

$$\tan 35° = \frac{h}{90}$$

$$h = 90 \times \tan 35°$$

$$h = 90 \times 0.4738$$

$$h = 42.64\,feet$$

Thus, the height of the building is 42.64feet.

Trigonometry in Aviation

The aviation technology has been evolved in many up-gradations in the last few years. It has taken in account that the speed, direction, and distance as well as have to consider the speed and direction of the wind. The wind plays a vital role in when and how a flight will travel. This equation cab is solved by using trigonometry.

For example, if an airplane is traveling at 250 miles per hour, 55° of a north of east and the wind blowing due to south at 19 miles per hour. This calculation will be solved using the trigonometry and find the third side of the triangle that will lead the aircraft in the right direction.

Trigonometry in Criminology

Trigonometry is even used in the investigation of a crime scene. The functions of trigonometry are helpful to calculate a trajectory of a projectile and to estimate the causes of a collision in a car accident. Further, it is used to identify how an object falls or in what angle the gun is shot.

Trigonometry in Marine Biology

Trigonometry is often used by marine biologists for measurements to figure out the depth of sunlight that affects algae to photosynthesis. Using the trigonometric function and mathematical models, marine biologists estimate the size of larger animals like whales and also understand their behaviors.

Trigonometry in Navigation

Trigonometry is used in navigating directions; it estimates in what direction to place the compass to get a straight direction. With the help of a compass and trigonometric functions in navigation, it will help to pinpoint a location and also to find distance as well to see the horizon.

Other uses of Trigonometry

- The calculus is based on trigonometry and algebra.
- The fundamental trigonometric functions like sine and cosine are used to describe the sound and light waves.
- Trigonometry is used in oceanography to calculate heights of waves and tides in oceans.
- It used in the creation of maps
- It is used in satellite systems.

Trigonometry in Construction

In construction we need trigonometry to calculate the following:

- Measuring fields, lots and areas;
- Making walls parallel and perpendicular;
- Installing ceramic tiles;
- Roof inclination;

The height of the building, the width length etc. and the many other such things where it becomes necessary to use trigonometry.

Architects use trigonometry to calculate structural load, roof slopes, ground surfaces and many other aspects, including sun shading and light angles.

Trigonometry in Physics

In physics, trigonometry is used to find the components of vectors, model the mechanics of waves (both physical and electromagnetic) and oscillations, sum the strength of fields, and use dot and cross products. Even in projectile motion you have a lot of application of trigonometry.

Trigonometry in Marine Engineering

In marine engineering trigonometry is used to build and navigate marine vessels. To be more specific trigonometry is used to design the Marine ramp, which is a sloping surface to connect lower and higher level areas, it can be a slope or even a staircase depending on its application.

References

- Trigonometry, science: britannica.com, Retrieved 8 February, 2019
- Trigonometric-functions, maths: byjus.com, Retrieved 18 March, 2019
- InverseTrigonometricFunctions: wolfram.com, Retrieved 9 April, 2019
- Properties-inverse-functions, inverse-trigonometric-functions, maths: toppr.com, Retrieved 23 January, 2019
- Calculus: sosmath.com, Retrieved 15 June, 2019
- Law-of-sines: varsitytutors.com, Retrieved 8 August, 2019
- Law-of-cosines: varsitytutors.com, Retrieved 17 May, 2019
- Applications-of-trigonometry, maths: byjus.com, Retrieved 12 March, 2019
- Real-life-applications-of-trigonometry: embibe.com, Retrieved 24 July, 2019

5

Calculus

The mathematical study of continuous change is known as calculus. Differential calculus and integral calculus are the two main branches that fall under this domain. The convergence of infinite sequences and series to a well-defined limit are the common foundations of both branches. All the diverse principles of calculus have been carefully analysed in this chapter.

Calculus is a branch of mathematics that involves the study of rates of change. Before calculus was invented, all math was static: It could only help calculate objects that were perfectly still. But the universe is constantly moving and changing. No objects—from the stars in space to subatomic particles or cells in the body—are always at rest. Indeed, just about everything in the universe is constantly moving. Calculus helped to determine how particles, stars, and matter actually move and change in real time.

Calculus is used in a multitude of fields that you wouldn't ordinarily think would make use of its concepts. Among them are physics, engineering, economics, statistics, and medicine. Calculus is also used in such disparate areas as space travel, as well as determining how medications interact with the body, and even how to build safer structures

Differential Calculus

A branch of mathematics dealing with the concepts of derivative and differential and the manner of using them in the study of functions. Differential calculus is usually understood to mean classical differential calculus, which deals with real-valued functions of one or more real variables, but its modern definition may also include differential calculus in abstract spaces. Differential calculus is based on the concepts of real number; function; limit and continuity — highly important mathematical concepts, which were formulated and assigned their modern content during the development of mathematical analysis and during studies of its foundations. The central concepts of differential calculus — the derivative and the differential — and the apparatus developed in this connection furnish tools for the study of functions which locally look like linear functions or polynomials, and it is in fact such functions which are of interest, more than other functions, in applications.

Derivative

Let a function $y = f(x)$ be defined in some neighbourhood of a point $x0$. Let $\Delta x \neq 0$ denote the increment of the argument and let $\Delta y = f(x_0 + \Delta x) - f(x_0)$ denote the corresponding increment of the value of the function. If there exists a (finite or infinite) limit

$$\lim_{\Delta x \to 0} \frac{\Delta y}{\Delta x},$$

then this limit is said to be the derivative of the function f at x_0 it is denoted by $f'(x_0), \mathrm{df}(x_0)/dx, y', y'x, dy/dx$ Thus, by definition,

$$f'(x_0) = \lim_{\Delta x \to 0} \frac{\Delta y}{\Delta x} = \lim_{\Delta x \to 0} \frac{f(x_0 + \Delta x) - f(x_0)}{\Delta x}$$

The operation of calculating the derivative is called differentiation. If $f'(x_0)$ is finite, the function f is called differentiable at the point x_0 A function which is differentiable at each point of some interval is called differentiable in the interval.

Geometric Interpretation of the Derivative

Let C be the plane curve defined in an orthogonal coordinate system by the equation $y = f(x)$ where f is defined and is continuous in some interval J; let $M(x_0, y_0)$ be a fixed point on C, let $P(x, y)(x \in J)$ be an arbitrary point of the curve C and let MP be the secant (Fig. a). An oriented straight line MT (T a variable point with abscissa $x_0 + \Delta x$) is called the tangent to the curve C at the point M if the angle ϕ between the secant MP and the oriented straight line tends to zero as $x \to x_0$ (in other words, as the point $P \in C$ arbitrarily tends to the point M). If such a tangent exists, it is unique. Putting $x = x_0 + \Delta x, \Delta y = f(x_0 + \Delta x) - f(x_0)$ one obtains the equation $\tan \beta = \Delta y / \Delta x$ for the angle β between MP and the positive direction of the x – axis (Fig. a).

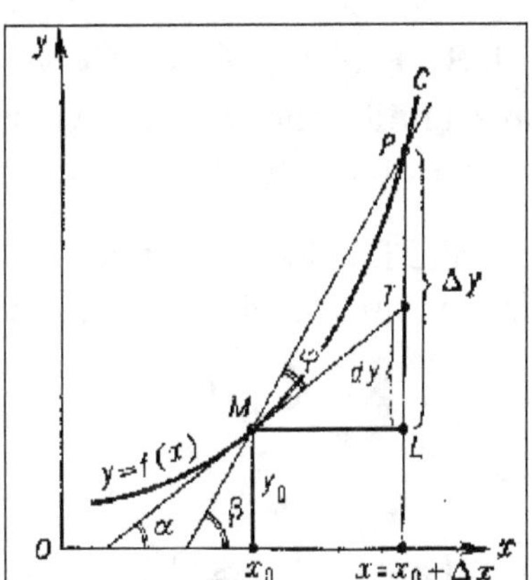

The curve C has a tangent at the point M if and only if $\lim_{\Delta x \to 0} \Delta y / \Delta x$ exists, i.e. if $f'(x_0)$ exists. The equation $\tan\alpha = f'(x_0)$ is valid for the angle α between the tangent and the positive direction of the x-axis. If $f'(x_0)$ is finite, the tangent forms an acute angle with the positive x-axis, i.e. $-\pi/2 < \alpha < \pi/2$; if $f'(x_0) = \infty$, the tangent forms a right angle with that axis.

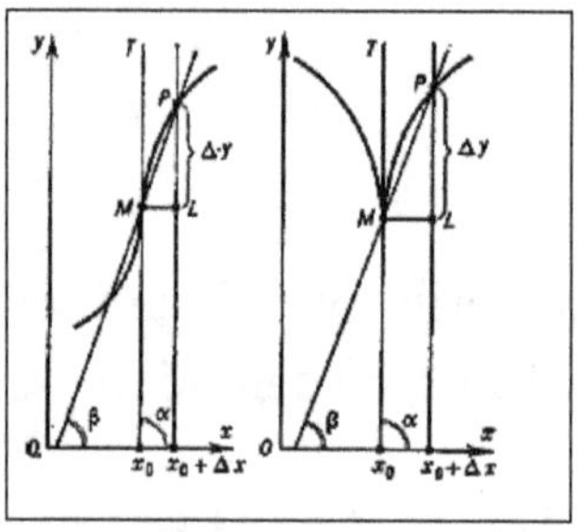

Thus, the derivative of a continuous function f at a point x_0 is identical to the slope $\tan\alpha$ of the tangent to the curve defined by the equation $y = f(x)$ at its point with abscissa x_0.

Mechanical interpretation of the derivative

Let a point M move in a straight line in accordance with the law $s = f(t)$. During time Δt the point M becomes displaced by $\Delta s = f(t+\Delta t) - f(t)$. The ratio $\Delta s/\Delta t$ represents the average velocity v_{av} during the time Δt. If the motion is non-uniform, V_{av} is not constant. The instantaneous velocity at the moment t is the limit of the average velocity as $\Delta t \to 0$, i.e. $v = f'(t)$ (on the assumption that this derivative in fact exists).

Thus, the concept of derivative constitutes the general solution of the problem of constructing tangents to plane curves, and of the problem of calculating the velocity of a rectilinear motion. These two problems served as the main motivation for formulating the concept of derivative.

A function which has a finite derivative at a point $x0$ is continuous at this point. A continuous function need not have a finite nor an infinite derivative. There exist continuous functions having no derivative at any point of their domain of definition.

The formulas given below are valid for the derivatives of the fundamental elementary functions at any point of their domain of definition (exceptions are stated):

1. If $f(x) = C = \text{const}$, then $f'(x) = C' = 0$;
2. If $f(x) = x$, then $f'(x) = 1$;
3. $(x^\alpha)' = \alpha x^{\alpha-1}$, $\alpha = \text{const}$ ($x \neq 0$, if $\alpha \leq 1$);

4. $(\alpha^x)' = \alpha^x$,In $a, a = \text{const} > 0$, $a \neq 1$; in particular, $(e^x)' = e^x$;
5. $(\log_\alpha x)' = (\log_\alpha e)/x = 1/(x\text{In}\quad)$, $a = \text{const} > 0, a \neq 1$. $(\text{In } x)' = 1/x$;
6. $(\sin x)' = \cos x$;
7. $(\cos x)' = -\sin x$;
8. $(\tan x)' = 1/\cos^2 x$;
9. $(\text{cotan } x)' = -1/\sin^2 x$;
10. $(\arcsin x)' = 1/\sqrt{1-x^2}, x \neq \pm 1$;
11. $(\arccos x)' = -1/\sqrt{1-x^2}, x \neq \pm 1$;
12. $(a\text{rxtan } x)' = -1/(1+x^2)$;
13. $(\text{arccotan } x)' = -1/(1+x^2)$;
14. $(\sinh x)' = \cosh x$;
15. $(\cosh x)' = \sinh x$;
16. $(\tanh x)' = 1/\cosh^2 x$;
17. $(\text{cotanh } x)' = -1/\sinh^2 x$.

The following laws of differentiation are valid:

If two functions u and v are differentiable at a point x_0, then the functions

$$cu \quad (\text{where } c = \text{const}), \quad u \pm v, \quad uv, \ \frac{u}{v}(v \neq 0)$$

are also differentiable at that point, and

$$(cu)' = cu',$$

$$(u \pm v)' = u' \pm v',$$

$$(uv)' = u'v + uv'$$

$$\left(\frac{u}{v}\right)' = \frac{u'v - uv'}{v^2}.$$

Theorem on the derivative of a composite function: If the function $y = f(u)$ is differentiable at a point u_0, while the function $\phi(x)$ is differentiable at a point x_0, and if $u_0 = \phi(x_0)$, then the composite function $y = f(\phi(x))$ is differentiable at x_0, and $y'x = f'(u_0)\phi'(x_0)$ or, using another notation, $dy/dx = (dy/du)(du/dx)$.

Theorem on the derivative of the inverse function: If $y = f(x)$ and $x = g(y)$ are two mutually inverse increasing (or decreasing) functions, defined on certain intervals, and if $f'(x_0) \neq 0$ exists (i.e. is not infinite), then at the point $y_0 = f(x_0)$ the derivative $g'(y_0) = 1/f'(x_0)$ exists, or, in a different notation, $dx/dy = 1/(dy/dx)$. This theorem may be extended: If the other conditions hold and if also $f'(x_0) = 0$ or $f'(x_0) = \infty$, then, respectively, $g'(y_0) = \infty$ or $g'(y_0) = 0$.

One-sided Derivatives

If at a point x_0 the limit,

$$\lim_{\Delta x \downarrow 0} \frac{\Delta y}{\Delta x}$$

exists, it is called the right-hand derivative of the function $y = f(x)$ at x_0 (in such a case the function need not be defined everywhere in a certain neighbourhood of the point x_0; this requirement may then be restricted to $x \geq x_0$). The left-hand derivative is defined in the same way, as:

$$\lim_{\Delta x \uparrow 0} \frac{\Delta y}{\Delta x}.$$

A function f has a derivative at a point x_0 if and only if equal right-hand and left-hand derivatives exist at that point. If the function is continuous, the existence of a right-hand (left-hand) derivative at a point is equivalent to the existence, at the corresponding point of its graph, of a right (left) one-sided semi-tangent with slope equal to the value of this one-sided derivative. Points at which the semi-tangents do not form a straight line are called angular points or cusps.

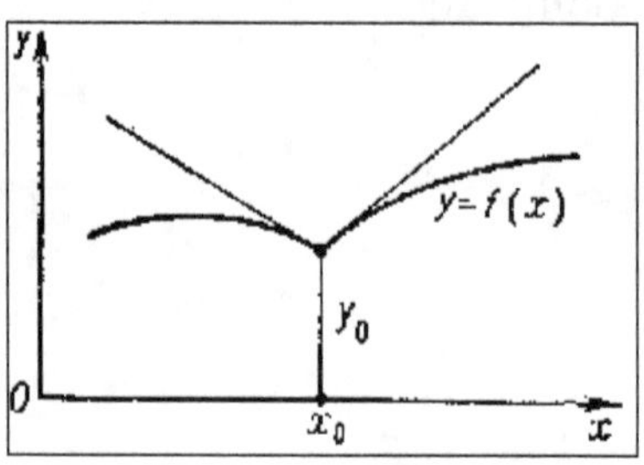

Derivatives of Higher Orders

Let a function $y = f(x)$ have a finite derivative $y' = f'(x)$ at all points of some interval; this derivative is also known as the first derivative, or the derivative of the first order, which, being a function of x, may in its turn have a derivative $y''=f''(x)$, known as the second derivative, or the derivative of the second order, of the function f, etc. In general, the n-th derivative, or the derivative of order n, is defined by induction by the equation $y^{\langle n\rangle}=\left(y^{\langle n-1\rangle}\right)'$, on the assumption that $y^{\langle n-1\rangle}$ is defined on some interval. The notations employed along with $y^{\langle n\rangle}$ are $f^{\langle n\rangle}$, $d^n f(x)/dx^n$, and, if $n=2,3.$, also y'', $f''(x)$, y''', $f'''(x)$.

The second derivative has a mechanical interpretation: It is the acceleration $w=d^2s/dt^2=f''(t)$ of a point in rectilinear motion according to the law $s=f(t)$.

Differential

Let a function $y=f(x)$ be defined in some neighbourhood of a point x and let there exist a number A such that the increment Δy may be represented as $\Delta y=A\Delta x+\omega$ with $\omega/\Delta x\to 0$ as $\Delta x\to 0$. The term $A\Delta x$ in this sum is denoted by the symbol dy or df and is named the differential of the function $f(x)$ (with respect to the variable x) at x. The differential is the principal linear part of increment of the function (its geometrical expression is the segment LT in Fig. , where MT is the tangent to $y=f(x)$ at the point (x_0, y_0) under consideration).

The function $y=f(x)$ has a differential at **x** if and only if it has a finite derivative,

$$f'(x)=\lim_{\Delta x\to 0}\frac{\Delta y}{\Delta x}=A$$

at this point. A function for which a differential exists is called differentiable at the point in question. Thus, the differentiability of a function implies the existence of both the differential and the finite derivative, and $dy=df(x)=f'(x)\Delta x$. For the independent variable x one puts $dx=\Delta x$, and one may accordingly write $dy=f'(x)dx$, i.e. the derivative is equal to the ratio of the differentials:

$$\Delta_x z=f(x_0+\Delta x, y_0)-f(x_0, y_0)$$

The formulas and the rules for computing derivatives lead to corresponding formulas and rules for calculating differentials. In particular, the theorem on the differential of a composite function is valid: If a function $y=f(u)$ is differentiable at a point u_0, while a function $\phi(x)$ is differentiable at a point x_0 and $u_0=\phi(x_0)$, then the composite function $y=f(\phi(x))$ is differentiable at the point x_0 and $dy=f'(u_0)du$, where $du=\phi'(x_0)dx$. The differential of a composite function has exactly the form it would have if the variable u were an independent variable. This property is known the invariance of the form of the differential. However, if u is an independent variable, $du=\Delta u$ is an arbitrary increment, but if u is a function, du is the differential of this function which, in general, is not identical with its increment.

Differentials of Higher Orders

The differential $d\,y$ is also known as the first differential, or differential of the first order. Let y = f (x) have a differential $d\,y = f'(x)\,dx$ at each point of some interval. Here dx =   is some number independent of and one may say, therefore that dx = const. The differential d y is a function of x alone, and may in turn have a differential, known as the second differential, or the differential of the second order, of *f*, etc. In general, the n -th differential, or the differential of order n, is defined by induction by the equality $d^n\,y = d(d^{n-1}y)$, on the assumption that the differential $d^{n-1}\,y$ is defined on some interval and that the value of dx is identical at all steps. The invariance condition for $d^2\,y, d^3\,y$ is generally not satisfied (with the exception y=f(u) where u is a linear function).The repeated differential of $d\,y$ has the form,

$$\delta(d\,y)\,f''(x)\;dx\delta x$$

and the value of $\delta(d\,y)$ for $dx = \delta x$ is the second differential.

Principal Theorems and Applications of Differential Calculus

The fundamental theorems of differential calculus for functions of a single variable are usually considered to include the Rolle theorem, the Legendre theorem (on finite variation), the Cauchy theorem, and the Taylor formula. These theorems underlie the most important applications of differential calculus to the study of properties of functions — such as increasing and decreasing functions, convex and concave graphs, finding the extrema, points of inflection, and the asymptotes of a graph. Differential calculus makes it possible to compute the limits of a function in many cases when this is not feasible by the simplest limit theorems. Differential calculus is extensively applied in many fields of mathematics, in particular in geometry.

Differential Calculus of Functions in Several Variables

For the sake of simplicity the case of functions in two variables (with certain exceptions) is considered below, but all relevant concepts are readily extended to functions in three or more variables. Let a function $z = f(x,y)$ be given in a certain neighbourhood of a point (xo, yo) and let the value $y = y_0$ be fixed. $f(x, y_0)$ will then be a function of alone. If it has a derivative with respect to x at xo, this derivative is called the partial derivative of f with respect to at ; it is denoted by $f'_x(x_0,y_0)\partial f(x_0,y_0)/\partial x\partial f/\partial x.z'_x\;\partial z/\partial x$ or $f_x(x_0,y_0)$ Thus, by definition,

$$f'_x(x_0,y_0)\lim_{\Delta y\to 0}\frac{\Delta_x z}{\Delta x} = \lim_{\Delta y\to 0}\frac{f(x_0,+\Delta x, x_0) - f(x_0,y_0)}{\Delta x},$$

where $\Delta_x z = f(x_0 + \Delta x, y_0) - f(x_0, y_0)$ is the partial increment of the function with respect to x (in the general case, $\partial z/\partial x$ must not be regarded as a fraction; $\partial/\partial x$ is the symbol of an operation).

The partial derivative with respect to y is defined in a similar manner:

$$f_y' (x_0,y_0)= \lim_{\Delta y\to 0}\frac{\Delta y\, z}{\Delta y}=\lim_{\Delta y\to 0}\frac{f(x_0,y_0+\Delta y)-f(x_0,y_0)}{\Delta y}$$

where $\Delta y\, z$ is the partial increment of the function with respect to y. Other notations include $\partial f(x_0,y_0)/\partial y \, \partial f/\partial y . z_y' . \partial z/\partial y$ *and* $f_y(x_0,y_0)$ Partial derivatives are calculated according to the rules of differentiation of functions of a single variable (in computing z_x' one assumes y=const while if $'$ is calculated, one assumes).

x=const

The partial differentials of $f(x,y)$ at (x_0,y_0) are, respectively,

$$d\ z=f'\ (x_0,y_0)\,dx;\quad d\ z=f'\ (x_0,y_0)\,d\,y$$

where, as in the case of a single variable, $dx=\Delta x, d\,y=\Delta y$, denote the increments of the independent variables.

The first partial derivatives $\partial z/\partial x=f_x'(x,y)$ *and* $\partial z/\partial y=f_y'(x_0,y_0)$, or the partial derivatives of the first order, are functions of x and y, and may in their turn have partial derivatives with respect to x and y. These are named, with respect to the function $z=f(x,y)$, the partial derivatives of the second order, or second partial derivatives. It is assumed that,

$$\frac{\partial}{\partial x}=\left(\frac{\partial z}{\partial x}\right)=\frac{\partial^2 z}{\partial x^2},\quad \frac{\partial}{\partial y}\left(\frac{\partial z}{\partial x}\right)=\frac{\partial^2 z}{\partial x\,\partial y}$$

$$\frac{\partial}{\partial x}=\left(\frac{\partial z}{\partial y}\right)=\frac{\partial^2 z}{\partial y\partial x},\quad \frac{\partial}{\partial y}\left(\frac{\partial z}{\partial y}\right)=\frac{\partial^2 z}{\partial y^2}.$$

The following notations are also used instead of $\partial^2 z/\partial x^2$.

$$z_{xx}''\quad z_x''2, \frac{\partial^2 f(x,y)}{\partial x^2}, \frac{\partial^2 f}{\partial x^2}, f_{xx}''\ (x,y)\, f\ _x''2(x,y), f_{xx}(x,y);$$

and instead of $\partial^2 z/\partial x\partial y$:

$$z_{xy}''\ , \frac{\partial^2 f(x,y)}{\partial x\partial y}, \frac{\partial^2 f}{\partial x\partial y}, f_{xy}''\ (x,y)\, f_{xy}(x,y),$$

etc. One can introduce in the same manner partial derivatives of the third and higher orders, together with the respective notations: $\partial^n z/\partial x^n$ means that the function z is to be differentiated n times with respect to x. $\partial^n z/\partial x^p\,\partial y^q$ where $n=p+q$ means that the function is differentiated p times with respect to x and q times with respect to y. The partial derivatives of second and higher orders obtained by differentiation with respect to different variables are known as mixed partial derivatives.

To each partial derivative corresponds some partial differential, obtained by its multiplication by the differentials of the independent variables taken to the powers equal to the number of differentiations with respect to the respective variable? In this way one obtains the -th partial differentials, or the partial differentials of order: n

$$\frac{\partial^n z}{\partial x^n} dx^n, \quad \frac{\partial^n z}{\partial x^p \partial y^q} dx^p \, dy^q.$$

The following important theorem on derivatives is valid: If, in a certain neighbourhood of a point (x_0, y_0), a function $z = f(x, y)$ has mixed partial derivatives $f\,x''y\,(x, y)$ and $f\,y''x(x, y)$, and if these derivatives are continuous at the point (x_0, y_0), then they coincide at this point.

A function $z = f(x, y)$ is called differentiable at a point (x_0, y_0) with respect to both variables x and y if it is defined in some neighbourhood of this point, and if it's total increment

$$\Delta z = f(x_0 + \Delta x, y_0 + \Delta y) - f(x_0, y_0)$$

may be represented in the form,

$$\Delta z = A\ \Delta x + B \Delta y + \ + \omega$$

Where A and B are certain numbers and $\omega / \rho \to 0$ for $\rho = \sqrt{(\Delta x)^2 + (\Delta y)^2} \to 0$ (provided that the point $(x_0 + \Delta x, y_0 + \Delta y)$ lies in this neighbourhood). In this context, the expression,

$$dz = d\,f(x_0, y_0) = A\,\Delta x + B\,\Delta y$$

is called the total differential (of the first order) of f' at (x_0, y_0) this is the principal linear part of increment. A function which is differentiable at a point is continuous at that point (the converse proposition is not always true!). Moreover, differentiability entails the existence of finite partial derivatives,

$$f'_x(x_0, y_0) = \lim_{\Delta x \to 0} \frac{\Delta x\, z}{\Delta x} = A, \quad f'_y(x_0, y_0) = \lim_{\Delta y \to 0} \frac{\Delta y\, z}{\Delta y} = B,$$

Thus, for a function which is differentiable at, (x_0, y_0)

$$dz = d\,f(x_0, y_0) = f'_x(x_0, y_0)\Delta x + f'_y(x_0, y_0)\Delta y,$$

or

$$dz = d\,f(x_0, y_0) = f'_x(x_0, y_0)\,dx + f'_y(x_0, y_0)\,dy,$$

if, as in the case of a single variable, one puts, for the independent variables, $dx = \Delta x, dy = \Delta y$.

The existence of finite partial derivatives does not, in the general case, entail differentiability (unlike in the case of functions in a single variable). The following is a sufficient criterion of the differentiability of a function in two variables: If, in a certain neighbourhood of a point (x_0, y_0), a function f has finite partial derivatives f'_x and f'_y which are continuous at, then is differentiable at this point. Geometrically, the total differential $df(x_0, y_0)$ is the increment of the applicate of the tangent plane to the surface $z = f(x, y)$ at the point (x_0, y_0, z_0), where $z_0 = f(x_0, y_0)$.

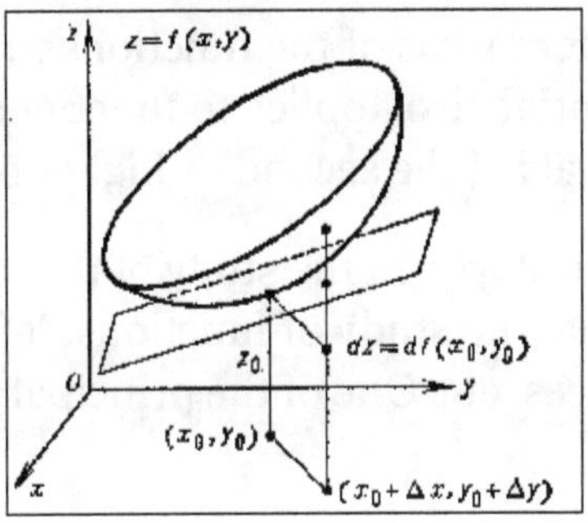

Total differentials of higher orders are, as in the case of functions of one variable, introduced by induction, by the equation,

$$d^n z = d(d^{n-1} z),$$

on the assumption that the differential $d^{n-1} z$ is defined in some neighbourhood of the point under consideration, and that equal increments of the dx, dy arguments, are taken at all steps. Repeated differentials are defined in a similar manner.

Derivatives and Differentials of Composite Functions

Let $w = f(u_1, ..., u_m)$ be a function m in variables which is differentiable at each point of an open D domain of the $m-$dimensional Euclidean space R^m, and let m functions $u_1 = \phi_1(x_1, ..., x_n), ..., u_m = \phi m(x_1, ..., x_n)$ in n variables be defined in an open domain G of the n -dimensional Euclidean space R^n . Finally, let the point $(u_1, ..., u_m)$, corresponding to a point $(x_1, ..., x_n) \in G$, be contained in . D The following theorems then hold:

A) If the functions $\phi_1, ..., \phi_m$ have finite partial derivatives with respect to $x_1, ..., x_n$, the composite function $w = f(u_1, ..., u_m)$ in $x_1, ..., x_n$ also has finite partial derivatives with respect to $x_1, ..., x_n$, and

$$\frac{\partial w}{\partial x_1} = \frac{\partial f}{\partial u_1}\frac{\partial u_1}{\partial x_1} + ... + \frac{\partial f}{\partial u_n}\frac{\partial u_n}{\partial x_1},$$

$$\frac{\partial w}{\partial x_n}=\frac{\partial f}{\partial u_1}\frac{\partial u_1}{\partial x_n}+\ldots+\frac{\partial f}{\partial u_n}\frac{\partial u_n}{\partial x_n},$$

B) If the functions $\phi_1,\ldots,\phi_m$ are differentiable with respect to all variables at a point $(x_1,\ldots,x_n)\in G$, then the composite function $w=f(u_1,\ldots,u_m)$ is also differentiable at that point, and

$$dw=\frac{\partial f}{\partial u_1}du_1+\ldots+\frac{\partial f}{\partial u_n}du_n,$$

Where $du_1,\ldots,du_m$ are the differentials of the functions . $u_1,\ldots,u_m$ Thus, the property of invariance of the first differential also applies to functions in several variables. It does not usually apply to differentials of the second or higher orders.

Differential calculus is also employed in the study of the properties of functions in several variables: finding extrema, the study of functions defined by one or more implicit equations, the theory of surfaces, etc. One of the principal tools for such purposes is the Taylor formula.

The concepts of derivative and differential and their simplest properties, connected with arithmetical operations over functions and superposition of functions, including the property of invariance of the first differential, are extended, practically unchanged, to complex-valued functions in one or more variables, to real-valued and complex-valued vector functions in one or several real variables, and to complex-valued functions and vector functions in one or several complex variables. In functional analysis the ideas of the derivative and the differential are extended to functions of the points in an abstract space.

Limits

Suppose we have a function f(x). The value, a function attains, as the variable x approaches a particular value say a, i.e., x → a is called its limit. Here, 'a' is some pre-assigned value. It is denoted as

$$\lim_{x\to a} f(x)=1$$

- The expected value of the function shown by the points to the left of a point 'a' is the left-hand limit of the function at that point. It is denoted as $\lim_{x\to a^-} f(x)$.
- The points to the right of a point 'a' which shows the value of the function is the right-hand limit of the function at that point. It is denoted as $\lim_{x\to a^+} f(x)$.

Limits of functions at a point are the common and coincidence value of the left and right-hand limits.

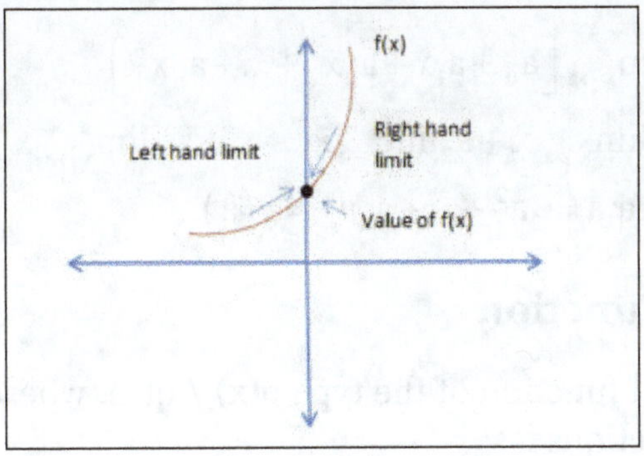

The value of a limit of a function f(x) at a point a i.e., f(a) may vary from the value of f(x) at 'a'.

Algebra of Limits

Let p and q be two functions such that their limits $\lim_{x\to a} p(x)$ and $\lim_{x\to a} q(x)$ exist.

- Limit of the sum of two functions is the sum of the limits of the functions.

$$\lim_{x\to a}\left[p(x)+q(x)\right]=\lim_{x\to a} p(x)+\lim_{x\to a} q(x).$$

- Limit of the difference of two functions is the difference of the limits of the functions.

$$\lim_{x\to a}\left[p(x)-q(x)\right]=\lim_{x\to a} p(x)-\lim_{x\to a} q(x).$$

- Limit of product of two functions is the product of the limits of the functions.

$$\lim_{x\to a}\left[p(x)\times q(x)\right]=\left[\lim_{x\to a} p(x)\right]\times\left[\lim_{x\to a} q(x)\right].$$

- Limit of quotient of two functions is the quotient of the limits of the functions.

$$\lim_{x\to a}\left[p(x)\div q(x)\right]=\left[\lim_{x\to a} p(x)\right]\div\left[\lim_{x\to a} q(x)\right].$$

- Limit of product of a function p(x) with a constant, q(x) = α is α times the limit of p(x).

$$\lim_{x\to a}\left[\alpha.p(x))\right]=\alpha.\lim_{x\to a} p(x).$$

Limit of Polynomial Function

Consider a polynomial function, $f(x)=a_0+a_1x+a_2x^2+\ldots+a_nx^n$. Here, $a_0, a_1, \ldots, a_n$ are all constants. At any point x = a, the limit of this polynomial function is

$$\lim_{x\to a} f(x) = \lim_{x\to a}\left[a_0 + a_1 x + a_2 x^2 + \ldots + a_n x^n\right]$$
$$= \lim_{x\to a} a_0 + a_1 \lim_{x\to a} x + a_2 \lim_{x\to a} x^2 + \ldots + a_n \lim_{x\to a} x^n$$
$$\text{or, } \lim_{x\to a} = a_0 + a_1 a + a_2 a^2 + \ldots + a_n a^n = f(a).$$

Limit of Rational Function

The limit of any rational function of the type p(x) / q(x), where q(x) ≠ o and p(x) and q(x) are polynomial functions is,

$$\lim_{x\to a}\left[p(x)/q(x)\right] = \left[\lim_{x\to a} p(x)\right]/\left[\lim_{x\to a} q(x)\right] = p(a)/q(a).$$

The very first step to find the limit of a rational function is to check if it is reduced to the form o/o at some point. If it is so, then we need to do some adjustment so that one can calculate the value of the limit. This can be done by,

- Canceling the factor which causes the limit to be of the form o/o.

Assume a function, $f(x) = (x^2 + 4x + 4)/(x^2 - 4)$. Taking limit over it for $x = -2$, the function is of the form o/o,

$$\lim_{x\to -2} f(x) = \lim_{x\to -2}\left[(x^2 + 4x + 4)/(x^2 - 4)\right]$$
$$= \lim_{x\to -2}\left[(x+2)^2/(x-2)(x+2)\right] = \lim_{x\to -2}(x+2)/(x-2) = 0/-4(\neq 0/0) = 0.$$

- Applying the L – Hospital's Rule.

Differentiating both the numerator and the denominator of the rational function until the value of limit is not of the form o/o. Assume a function, f(x) = sin x/x. Taking limit over it for x = o, the function is of the form o/o.

Taking the differentiation of both sin x and x with respect to x in the limit $\lim_{x\to 0}$ sinx/x reduces to $\lim_{x\to 0}$ cos x/1=1. $(\cos 0 = 1)$.

One-Sided Limits

In order to calculate a limit at a point, we need to have an interval around that point; that is, we consider values of the function for x values on both sides of the point. Since we are considering values on both sides of the point, this type of limit is sometimes referred to as a two-sided limit. At some points, such as end points, it is not possible to find an interval on both sides of the point; for endpoints we can only find an interval on one side of the point. Instead, we can use the information that we are provided on that interval, in order to calculate a one-sided limit. In this way, we can define left-hand and right-hand limits, looking at the function from the left or right side of the point, respectively. We write the left-hand limit of $f(x)$, or the limit as x approaches x_0 from the left-hand side as,

$$\lim_{x\to x_0^-} f(x)$$

and we write the right-hand limit as,

$$\lim_{x\to x_0^+} f(x)$$

where the – and + denote whether it is approaching from the left or right hand side, respectively.

More formally, we have the following definitions.

Right-hand Limit

We say that L is the right-hand limit of $f(x)$ at x_0, written,

$$\lim_{x\to x_0^+} f(x) = L$$

if for every number $\varepsilon > 0$, there exists a corresponding number $\delta > 0$ such that for all x,

$$x_0 < x < x_0 + \delta \Rightarrow |f(x) - L| < \varepsilon$$

Left-hand Limit

We say that L is the left-hand limit of $f(x)$ at x_0, written,

$$\lim_{x\to x_0^-} f(x) = L$$

if for every number $\varepsilon > 0$, there exists a corresponding number $\delta > 0$ such that for all x,

$$x_0 - \delta < x < x_0 \Rightarrow |f(x) - L| < \epsilon$$

It is noteworthy that all of the rules for combining two-sided limits also apply for combining one-sided limits.

Example: Find $\lim_{x\to 0^+} f(x)$ and $\lim_{x\to 0^-} f(x)$ for $f(x) = \frac{|x|}{x}$.

Solution The solution to this problem becomes much more evident if we rewrite $f(x)$ as-

$$f(x)=\begin{cases}-1 & x<0\\ 1 & x>0\end{cases}$$

Now we can see that looking from just the left or right side of the point $x\ =0$, we have two constant functions. Since the limit of a constant is just that constant, it follows that,

$$\lim_{x\to 0^+} f(x)=1 \text{ and } \lim_{x\to 0^-} f(x)=-1$$

The following theorem is a useful tool for relating one-sided and two-sided limits.

Theorem: One-sided and Two-sided Limits

A function $f(x)$ has a limit L at x_0 if and only if it has right-hand and left-hand limits at x_0, and both of those limits are L.

If both of the one-sided limits have the same value L, then we can certainly construct a δ-interval on both sides of x_0 by combining both of the one-sided intervals, which implies the two-sided limit exists. If the one-sided limits exist but disagree, then it is impossible for the function to approach a single value as $x\ \rightarrow\ x_0$,, which implies that the two-sided limit does not exist. From this we can conclude that $\lim_{x\to 0}\dfrac{|x|}{x}$ does not exist. This is a much more efficient way to prove a limit does not exist than proving that it does not exist for all possible values L.

Example: Prove that

$$\lim_{x\to 0^+} \sqrt{x}=0$$

Solution: Consider $\varepsilon\ >\ 0$,, arbitrary. We need to find $\delta\ >\ 0$ so that for all x with $2\ -\ \delta<x<2$ we have $\left|\sqrt{x}-0\right|<\varepsilon$ or $\sqrt{x}<\varepsilon$. Manipulating this inequality,

$$\sqrt{x}<\varepsilon$$

$$0\le x<\varepsilon^2$$

Thus, if we set $\delta\ =\ \varepsilon^2$, for any x with $0<x<\delta=\varepsilon^2$ we have:

$$\sqrt{x}<\sqrt{\varepsilon^2}=\varepsilon$$

and the conclusion follows.

Example: Let $f(x)$ be given by,

$$f(x)=\sqrt{4-x^2}$$

Find the one-sided limits at the endpoints of the domain of this function. Using the definition of left and right-hand limits, prove that these limits exist, for some values L.

Solution First we must recall that $\sqrt{x}$ is not definied on $\mathbb{R}$ for $x<0$. In this way, we can determine that if $|x|>2$ then $f(x)=\sqrt{4-x^2}$ is not defined. Thus, we can see that the domain of this function is [−2, 2]. On this domain our function is a semicircle. At the left endpoint we must consider the right-hand limit, and at the right endpoint we consider the left-hand limit. Using the rules for combining limits,

$$\lim_{x\to -2^+}\sqrt{4-x^2}=0 \text{ and } \lim_{x\to -2^-}\sqrt{4-x^2}=0$$

Now our task is to prove that these limits exist as written above, using the definition of one-sided limits. We will prove that the limit as $x\to 2^-$ is 0, and leave the analagous proof at the left endpoint to the reader.

Consider $\varepsilon > 0$, arbitrary. We need to find a $\delta > 0$ so that for all x with $2 - \delta < x < 2$ we have,

$$\left|\sqrt{4-x^2}-0\right|<\varepsilon$$

In this problem it will be difficult to directly manipulate the second inequality in order to find a sufficiently small value for δ. We will need to take a slightly more creative approach. Notice that,

$$\sqrt{4-x^2}$$

will be at its largest when x is the smallest, or when x is at the farthest left point of the interval $2-\delta<x<2$. Thus, if we can find a value for δ such that,

$$\sqrt{4-(2-\delta)^2}<\varepsilon$$

we will have a δ such that for all x in the interval $2-\delta<x<2$ the function values are within the error tolerance ε of 0. Thus, for all x in the interval,

$$\sqrt{4-x^2}<\sqrt{4-(2-\delta)^2}=\sqrt{4-\left(4-4\delta+\delta^2\right)}=\sqrt{4-4+4\delta-\delta^2}=\sqrt{4\delta-\delta^2}=\sqrt{\delta(4-\delta)}$$

The first thing to note is that because we cannot have a negative input to $\sqrt{\delta(4-\delta)}$ we need to have $(4-\delta)\geq 0$ which means that,

$$\delta\leq 4$$

Now, to have

$$\sqrt{\delta(4-\delta)} < \varepsilon$$

we need,

$$\delta(4-\delta) < 4\delta < \varepsilon^2$$

From this inequality, we obtain the restriction,

$$\delta < \frac{\varepsilon^2}{4}$$

Thus, we set-

$$\delta = \min\left(\frac{\in^2}{16}, 1\right)$$

in order to encapsulate both of the previous restrictions on δ we found (there is nothing unique about the values we chose, just that they satisfy $\delta \le 4$ and $\delta < \in^2/4)$. Now consider arbitrary x with $2-\delta < x < 2$. It follows:

$$\sqrt{4-x^2} < \sqrt{4-(2-\delta)^2} = \sqrt{\delta(4-\delta)} < \sqrt{\frac{\in^2}{16}.4} = \frac{\in}{2} < \in$$

After this long and arduous analysis, we have managed to prove the one-sided limit exists, and is equal to 0.

Properties of Limits

Sum Rule

This rule states that the limit of the sum of two functions is equal to the sum of their limits:

$$\lim_{x\to a}\left[f(x)+g(x)\right] = \lim_{x\to a} f(x) + \lim_{x\to a} g(x).$$

Extended Sum Rule

$$\lim_{x\to a}\left[f_1(x)+\ldots+f_n(x)\right] = \lim_{x\to a} f_1(x)+\ldots+\lim_{x\to a} f_n(x).$$

Constant Function Rule

The limit of a constant function is the constant:

$$\lim_{x\to a} C = C.$$

Constant Multiple Rule

The limit of a constant times a function is equal to the product of the constant and the limit of the function:

$$\lim_{x\to a} k\, f(x) = k \lim_{x\to a} f(x).$$

Product Rule

This rule says that the limit of the product of two functions is the product of their limits (if they exist):

$$\lim_{x\to a}\left[f(x)g(x)\right] = \lim_{x\to a} f(x).\lim_{x\to a} g(x).$$

Extended Product Rule

$$\lim_{x\to a}\left[f_1(x)f_2(x)...f_n(x)\right] = \lim_{x\to a} f_1(x).\lim_{x\to a} f_2(x)...\lim_{x\to a} f_n(x).$$

Quotient Rule

The limit of quotient of two functions is the quotient of their limits, provided that the limit in the denominator function is not zero:

$$\lim_{x\to a}\frac{f(x)}{f(x)} = \frac{\lim\limits_{x\to a} f(x)}{\lim\limits_{x\to a} g(x)},\ \text{if } \lim_{x\to a} g(x) \neq 0.$$

Power Rule

$$\lim_{x\to a}\left[f(x)\right]^P = \left[\lim_{x\to a} f(x)\right]^P,$$

where the power p can be any real number. In particular,

$$\lim_{x\to a}\sqrt[p]{f(x)} = \sqrt[p]{\lim_{x\to a} f(x)}.$$

If $\mathrm{f}(x) = x^n$, then

$$\lim_{x\to a} x^n = a^n, n = 0 \pm 1, \pm\ 2, ...\ and\ a \neq 0, \text{ if } n \leq 0.$$

This is a special case of the previous property.

Limit of an Exponential Function

$$\lim_{x \to a} b^{f(x)} = b^{\lim_{x \to a} f(x)},$$

where the base b > 0.

Limit of a Logarithm of a Function

$$\lim_{x \to a} \left[\log_b f(x)\right] = \log_b \left[\lim_{x \to a} f(x)\right],$$

where the base b > 0.

Integral Calculus

Integral calculus is the branch of mathematics in which the notion of an integral, its properties and methods of calculation are studied. Integral calculus is intimately related to differential calculus, and together with it constitutes the foundation of mathematical analysis.

By means of integral calculus it became possible to solve by a unified method many theoretical and applied problems, both new ones which earlier had not been amenable to solution, and old ones that had previously required special artificial techniques. The basic notions of integral calculus are two closely related notions of the integral, namely the indefinite and the definite integral.

The indefinite integral of a given real-valued function on an interval on the real axis is defined as the collection of all its primitives on that interval, that is, functions whose derivatives are the given function. The indefinite integral of a function f is denoted by $\int f(x)\,dx$. If F is some primitive of f, then any other primitive of it has the form $F + C$, where C is an arbitrary constant; one therefore writes,

$$\int f(x)\,dx = F(x) + C.$$

The operation of finding an indefinite integral is called integration. Integration is the operation inverse to that of differentiation:

$$\int F'(x)\,dx = F(x) + C, \quad d\int f(x)\,dx = f(x)\,dx.$$

The operation of integration is linear: If on some interval the indefinite integrals,

$$\int f_1(x)\,dx \ \ and \ \int f_2(x)\,dx$$

exist, then for any real numbers and, the following integral exists on this interval:

$$\int [\lambda_1 f_1(x) + \lambda_2 f_2(x)] dx$$

and equals-

$$\lambda_1 \int f_1(x) dx + \lambda_2 \int f_2(x) dx.$$

For indefinite integrals, the formula of integration by parts holds: If two functions and are differentiable on some interval and if the integral exists, then so does the integral, and the following formula holds:

$$\int u dv = uv - \int v du.$$

The formula for change of variables holds: If for two functions f and ϕ defined on certain intervals, the composite function $f o \phi$ makes sense and the function ϕ is differentiable, then the integral,

$$\int f[\phi(t)] \phi'(t) dt$$

exists and equals-

$$\int f(x) dx.$$

A function that is continuous on some bounded interval has a primitive on it and hence an indefinite integral exists for it. The problem of actually finding the indefinite integral of a specified function is complicated by the fact that the indefinite integral of an elementary function is not an elementary function, in general. Many classes of functions are known for which it proves possible to express their indefinite integrals in terms of elementary functions. The simplest examples of these are integrals that are obtained from a table of derivatives of the basic elementary functions:

1. $\int x^{\alpha} dx = \frac{x^{\alpha+1}}{\alpha+1} + C, \alpha \neq -1;$
2. $\int \frac{dx}{x} = \text{In } |x| + C;$
3. $\int a^x dx = \frac{a^x}{In\ a} + C. a > 0, a \neq 1;$ in particular $\int e^x dx = e^x + C;$
4. $\int \sin x dx = -\cos x + C;$
5. $\int \cos x dx = -\sin x + C;$

6. $\int \frac{dx}{\cos^2 x} = \tan x + C;$

7. $\int \frac{dx}{\sin^2 x} = -\mathrm{cotan}\, x + C;$

8. $\int \sinh x\, dx = \cosh x + C;$

9. $\int \cosh x\, dx = \sinh x + C;$

10. $\int \frac{dx}{\cosh^2 x} = \tanh x + C;$

11. $\int \frac{dx}{\sinh^2 x} = -\mathrm{cotah}\, x + C;$

12. $\int \frac{dx}{x^2 + a^2} = \frac{1}{a} \arctan \frac{x}{a} + C = -\frac{1}{a}\ \mathrm{arccotan}\ \frac{x}{a} + C';$

13. $\int \frac{dx}{x^2 - a^2} = \frac{1}{2a} In \left| \frac{x-a}{x+a} \right| + C;$

14. $\int \frac{dx}{\sqrt{a^2\ \ x^2}} = \arcsin \frac{x}{a} + C = -\arccos \frac{x}{a} + C', |x|\ |a|$

15. $\int \frac{dx}{\sqrt{x^2 \pm a^2}} = \mathrm{In}\ \left| x + \sqrt{x^2 \pm a^2} \right| + C$ (when $x^2 - a^2$ is under the square root, it is assumed that $(|x| > |a|)$.).

If the denominator of the integrand vanishes at some point, then these formulas are valid only for those intervals inside which the denominator does not vanish.

The indefinite integral of a rational function over any interval on which the denominator does not vanish is a composition of rational functions, arctangents and natural logarithms. Finding the algebraic part of the indefinite integral of a rational function can be achieved by the Ostrogradski method. Integrals of the following types can be reduced by means of substitution and integration by parts to integration of rational functions:

$$\int R\left[x, \left(\frac{ax+b}{cx+b}\right)^{r_1}, \ldots, \left(\frac{ax+b}{cx+b}\right)^{r_m}\right] dx,$$

where $r1, \ldots . rm$ are rational numbers; integrals of the form,

$$\int R\left(x\sqrt{ax^2 + bx + c}\right) dx$$

Certain cases of integrals of differential binomials; integrals of the form

$$\int R(\sin x, \cos x)\,dx, \quad \int R(\sinh, x\cosh x)\,dx$$

(where $R(y1,.....yn)$ are rational functions); the integrals,

$$\int e^{\alpha x}\cos\beta x dx, \quad \int e^{\alpha x}\sin\beta x dx,$$

$$\int x^n \cos\alpha\, x dx, \quad \int x^n \sin\alpha\, x dx,$$

$$\int x^n \arcsin x dx, \quad \int x^n \arccos x dx,$$

$$\int x^n \arctan x dx, \quad \int x^n \operatorname{arccotan} x dx, \quad n = 0,1,...,$$

and many others. In contrast, for example, the integrals

$$\int \frac{e^x}{x^n}dx, \int \frac{\sin x}{x^n}dx, \int \frac{\cos x}{x^n}dx, \quad n = 1,2,....,$$

cannot be expressed in terms of elementary functions.

The definite integral

$$\int_a^b f(x)\,dx$$

of a function f defined on an interval $[a,b]$ is the limit of integral sums of a specific type. If this limit exists, f is said to be Cauchy, Riemann, Lebesgue, etc. integrable.

The geometrical meaning of the integral is tied up with the notion of area: If the function $f \geq 0$ is continuous on the interval $[a,b]$, then the value of the integral

$$\int_a^b f(x)\,dx$$

is equal to the area of the curvilinear trapezium formed by the graph of the function, that is, the set whose boundary consists of the graph of f, the segment $[a,b]$ and the two segments on the lines $x = a$ and $x = b$ making the figure closed, which may degenerate to points.

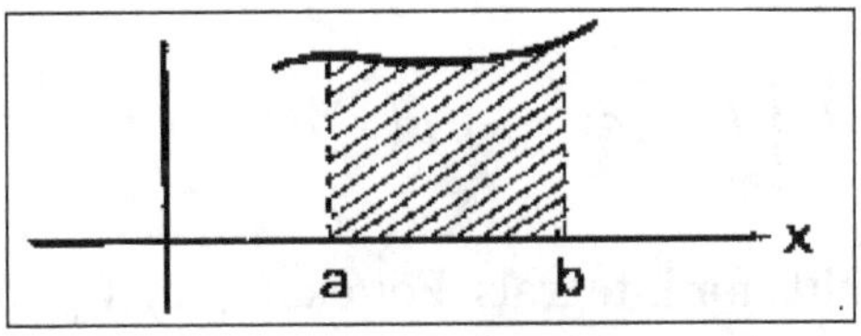

The calculation of many quantities encountered in practice reduces to the problem of calculating the limit of integral sums; in other words, finding a definite integral; for example, areas of figures and surfaces, volumes of bodies, work done by force, the coordinates of the centre of gravity, the values of the moments of inertia of various bodies, etc.

The definite integral is linear: If two functions f_1 and f_2 are integrable on an interval $[a,b]$, then for any real numbers λ_1 and λ_2 the function,

$$\lambda_1 f_1 + \lambda_2 f_2$$

is also integrable on this interval and,

$$\int_a^b [\lambda_1 f_1(x) + \lambda_2 f_2(x)]dx = \lambda_1 \int_a^b f_1(x)dx + \lambda_2 \int_a^b f_2(x)dx.$$

Integration of a function over an interval has the property of monotonicity: If the function f is integrable on the interval $[a\ b]$ and if $[c,d] \subset [a,b]$, then f is integrable on $[c,b]$ as well. The integral is also additive with respect to the intervals over which the integration is carried out: If $a < c < b$ and the function f is integrable on the intervals $[a,b]$ and $[c,b]$, then it is integrable on $[a,b]$, and

$$\int_a^b f(x)dx = \int_a^c f(x)dx + \int_c^b f(x)dx.$$

If f and g are Riemann integrable, then their product is also Riemann integrable. If $f \geq g$ on $[a,b]$, then

$$\int_a^b f(x)dx \geq \int_a^b g(x)dx.$$

If f is integrable on $[a,b]$, then the absolute value $|f|$ is also integrable on $[a,b]$ if $-\infty < a < b < \infty$,, and

$$\left|\int_a^b f(x)dx\right| \leq \int_a^b |f(x)|dx.$$

By definition one sets

$$\int_a^b f(x)dx = 0 \quad and \quad \int_b^a f(x)dx = -\int_a^b f(x)dx, \ a < 0.$$

A mean-value theorem holds for integrals. For example, if f and g are Riemann integrable on an interval $[a,b]$, if $m \leq f(x) \leq M, x \in [a,b]$, and if g does not change sign on [a, b], that is, it is either non-negative or non-positive throughout this interval, then there exists a number $m \le \mu \le M$ for which

$$\int_a^b f(x)g(x)dx = \mu \int_a^b g(x)dx.$$

Under the additional hypothesis that f is continuous on $[a,b]$, there exists in (a,b) a point ξ for which

$$\int_a^b f(x)g(x)dx = f(\xi)\int_a^b g(x)dx.$$

In particular, if $g(x) \equiv 1$, then

$$\int_a^b f(x)dx = f(\xi)(b-a).$$

Integrals with a Variable Upper Limit

If a function f is Riemann integrable on an interval $[a,b]$, then the function F defined by

$$F(x) = \int_a^x f(t)dt, \quad a \le x \le b,$$

is continuous on this interval. If, in addition, f is continuous at a point $x0$, then F is differentiable at this point and $F'(x_0) = f(x_0)$. In other words, at the points of continuity of a function the following formula holds:

$$\frac{d}{dx}\int_a^x f(t)dt = f(x).$$

Consequently, this formula holds for every Riemann-integrable function on an interval $[a,b]$, except perhaps at a set of points having Lebesgue measure zero, since if a function is Riemann integrable on some interval, then its set of points of discontinuity has measure zero. Thus, if the function f is continuous on $[a,b]$, then the function F defined by

$$F(x) = \int_a^x f(t)dt$$

is a primitive of f on this interval. This theorem shows that the operation of differentiation is inverse to that of taking the definite integral with a variable upper limit, and in this way a relationship is established between definite and indefinite integrals:

$$\int f(x)dx = \int_a^x f(t)dt + C.$$

The geometric meaning of this relationship is that the problem of finding the tangent to a curve and the calculation of the area of plane figures are inverse operations in the above sense.

The following Newton–Leibniz formula holds for any primitive F of an integrable function on an interval $[a,b]$:

$$\int_a^b f(x)dx = F(b) - F(a).$$

It shows that the definite integral of a continuous function over some interval is equal to the difference of the values at the end points of this interval of any primitive of it. This formula is sometimes taken as the definition of the definite integral. Then it is proved that the integral $\int_a^b f(x)dx$ introduced in this way is equal to the limit of the corresponding integral sums.

For definite integrals, the formulas for change of variables and integration by parts hold. Suppose, for example, that the function f is continuous on the interval (a,b) and that ϕ is continuous together with its derivative ϕ' on the interval (α,β), where (α,β) is mapped by ϕ into (a,b): $a < \phi(t) < b$ for $\alpha < t < \beta$,, so that the composite $f o \phi$ is meaningful in (α,β). Then, for $a_0, \beta_0 \in (\alpha,\beta)$,, the following formulas for change of variables holds:

$$\int_{\phi(a_0)}^{\phi(\beta_0)} f(x)dx = \int_{a_0}^{\beta_0} f\left[\phi(t)\phi'(t)dt.\right]$$

The formula for integration by parts is:

$$\int_a^b u(x)dv(x) = u(x)v(x)\Big|_{x=a}^{x=b} - \int_a^b v(x)du(x),$$

where the functions u and v have Riemann-integrable derivatives on $[a,b]$.

The Newton–Leibniz formula reduces the calculation of an indefinite integral to finding the values of its primitive. Since the problem of finding a primitive is intrinsically a difficult one, other methods of finding definite integrals are of great importance, among which one should mention the method of residues and the method of differentiation or integration with respect to the parameter of a parameter-dependent integral. Numerical methods for the approximate computation of integrals have also been developed.

Generalizing the notion of an integral to the case of unbounded functions and to the case of an unbounded interval leads to the notion of the improper integral, which is defined by yet one more limit transition.

The notions of the indefinite and the definite integral carry over to complex-valued functions. The representation of any holomorphic function of a complex variable in the form of a Cauchy integral over a contour played an important role in the development of the theory of analytic functions.

The generalization of the notion of the definite integral of a function of a single variable to the case of a function of several variables leads to the notion of a multiple integral.

For unbounded sets and unbounded functions of several variables, one is led to the notion of the improper integral, as in the one-dimensional case.

The extension of the practical applications of integral calculus necessitated the introduction of the notions of the curvilinear integral, i.e. the integral along a curve, the surface integral, i.e. the integral over a surface, and more generally, the integral over a manifold, which are reducible in some sense to a definite integral (the curvilinear integral reduces to an integral over an interval, the surface integral to an integral over a (plane) region, the integral over an n-dimensional manifold to an integral over an n-dimensional region). Integrals over manifolds, in particular curvilinear and surface integrals, play an important role in the integral calculus of functions of several variables; by this means a relationship is established between integration over a region and integration over its boundary or, in the general case, over a manifold and its boundary. This relationship is established by the Stokes formula, which is a generalization of the Newton–Leibniz formula to the multi-dimensional case.

Multiple, curvilinear and surface integrals find direct application in mathematical physics, particularly in field theory. Multiple integrals and concepts related to them are widely used in the solution of specific applied problems. The theory of cubature formulas has been developed for the numerical calculation of multiple integrals.

The theory and methods of integral calculus of real- or complex-valued functions of a finite number of real or complex variables carry over to more general objects. For example, the theory of integration of functions whose values lie in a normed linear space, functions defined on topological groups, generalized functions, and functions of an infinite number of variables (integrals over trajectories). Finally, a new direction in integral calculus is related to the emergence and development of constructive mathematics.

Integral calculus is applied in many branches of mathematics (in the theory of differential and integral equations, in probability theory and mathematical statistics, in the theory of optimal processes, etc.), and in applications of it.

Riemann Integral

The Riemann integral is the definite integral normally encountered in calculus texts and used by physicists and engineers. Other types of integrals exist (e.g., the Lebesgue integral), but are unlikely to be encountered outside the confines of advanced mathematics texts. In fact, according to Jeffreys and Jeffreys, "it appears that cases where these methods [i.e., generalizations of the Riemann integral] are applica-ble and Riemann's [definition of the integral] is not are too rare in physics to repay the extra difficulty."

The Riemann integral is based on the Jordan measure, and defined by taking a limit of a Riemann sum,

$$\int_a^b f(x)\,dx \equiv \lim_{\max \Delta x_k \to 0} \sum_{k=1}^{n} f(x_k^*)\,\Delta x_k$$

$$\iint f(x, y)\,dA \equiv \lim_{\max \Delta A_k \to 0} \sum_{k=1}^{n} f(x_k^*, y_k^*)\,\Delta A_k$$

$$\iiint f(x, y, z)\,dV \equiv \lim_{\max \Delta V_k \to 0} \sum_{k=1}^{n} f(x_k^*, y_k^*, z_k^*)\,\Delta V_k,$$

where $a \le x \le b$ and x_k^*, y_k^*, and z_k^* are arbitrary points in the intervals Δx_k, Δy_k, and Δz_k, respectively. The value $\max \Delta x_k$ is called the mesh size of a partition of the interval $[a, b]$ into subintervals Δx_k.

As an example of the application of the Riemann integral definition, find the area under the curve $y = x^r$ from 0 to a. Divide $(0, a)$ into n segments, so $\Delta x_k = a / n \equiv h$, then,

$$f(x_1) = f(0) = 0$$
$$f(x_2) = f(\Delta x_k) = h^r$$
$$f(x_3) = f(2\Delta x_k) = (2h)^r.$$

By induction,

$$f(x_k) = f([k-1]\Delta x_k) = [(k-1)h]^r = h^r (k-1)^r,$$

So,

$$f(x_k)\,\Delta x_k = h^{r+1} (k-1)^r$$

$$\sum_{k=1}^{n} f(x_k)\Delta x_k = h^{r+1} \sum_{k=1}^{n} (k-1)^r.$$

For example, take $r = 2$.

$$\sum_{k=1}^{n} f(x_k)\Delta x_k = h^3 \sum_{k=1}^{n} (k-1)^2$$
$$= h^3 \left(\sum_{k=1}^{n} k^2 - 2\sum_{k=1}^{n} k + \sum_{k=1}^{n} 1 \right)$$
$$= h^3 \left[\frac{n(n+1)(2n+1)}{6} - 2\frac{n(n+1)}{2} + n \right],$$

So,

$$I \equiv \lim_{n\to\infty} \sum_{k=1}^{n} f(x_k^*)\Delta x_k = \lim_{n\to\infty} \sum_{k=1}^{n} f(x_k)\Delta x_k$$
$$= \lim_{n\to\infty} h^3 \left[\frac{n(n+1)(2n+1)}{6} - 2\frac{n(n+1)}{2} + n \right]$$
$$= a^3 \lim_{n\to\infty} \left[\frac{n(n+1)(2n+1)}{6n^3} - \frac{n(n+1)}{n^3} + \frac{n}{n^3} \right]$$
$$= \frac{1}{3} a^3.$$

Lebesgue Integral

Lebesgue integral refers to the way of extending the concept of area inside a curve to include functions that do not have graphs representable pictorially. The graph of a function is defined as the set of all pairs of x- and y-values of the function. A graph can be represented pictorially if the function is piecewise continuous, which means that the interval over which it is defined can be divided into subintervals on which the function has no sudden jumps. Because the Riemann integral is based on the Riemann sums, which involve subintervals, a function not definable in this way will not be Riemann integrable.

For example, the function that equals 1 when x is rational and equals 0 when x is irrational has no interval in which it does not jump back and forth. Consequently, the Riemann sum,

$$f(c_1)\Delta x_1 + f(c_2)\Delta x_2 + \cdots + f(c_n)\Delta x_n$$

has no limit but can have different values depending upon where the points c are chosen from the subintervals Δ x.

Lebesgue sums are used to define the Lebesgue integral of a bounded function by partitioning the y-values instead of the x-values as is done with Riemann sums. Associated with the partition,

$$\{y_i\}\ (=y_0, y_1, y_2, \ldots, y_n)$$

are the sets E_i composed of all x-values for which the corresponding y-values of the function lie between the two successive y-values y_{i-1} and y_i. A number is associated with these sets E_i, written as $m(E_i)$ and called the measure of the set, which is simply its length when the set is composed of intervals. The following sums are then formed:

$$S = m(E_0)y_1 + m(E_1)y_2 + \cdots + m(E_{n-1})y_n$$

and

$$s = m(E_0)y_0 + m(E_1)y_1 + \cdots + m(E_{n-1})y_{n-1}.$$

As the subintervals in the y-partition approach 0, these two sums approach a common value that is defined as the Lebesgue integral of the function.

The Lebesgue integral is the concept of the measure of the sets E_i in the cases in which these sets are not composed of intervals, as in the rational/irrational function above, which allows the Lebesgue integral to be more general than the Riemann integral.

Contour Integrals

Contour integration is the process of calculating the values of a contour integral around a given contour in the complex plane. As a result of a truly amazing property of holomorphic functions, such integrals can be computed easily simply by summing the values of the complex residues inside the contour.

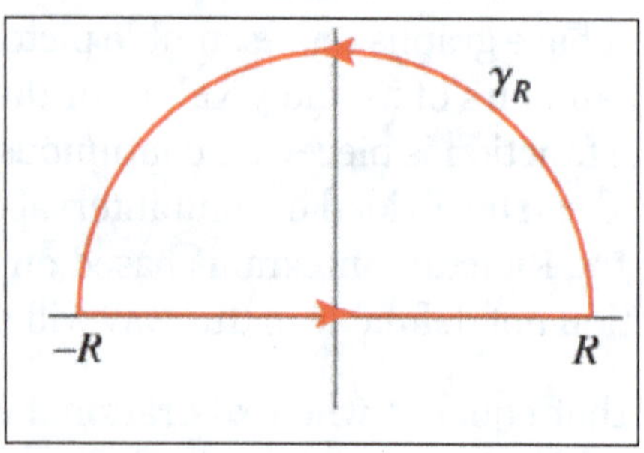

Let $P(x)$ and $Q(x)$ be polynomials of polynomial degree n and m with coefficients $b_n, \ldots, b_0$ and $c_m, \ldots, c_0$. Take the contour in the upper half-plane, replace x by z, and write $z \equiv R\, e^{i\theta}$. Then,

$$f, g, and\ h.$$

Define a path γ_R which is straight along the real axis from $-R$ to R and make a circular half-arc to connect the two ends in the upper half of the complex plane. The residue theorem then gives,

$$\lim_{R\to\infty}\int_{\gamma_R}\frac{P(z)dz}{Q(z)}=\lim_{R\to\infty}\int_{-R}^{R}\frac{P(z)dz}{Q(z)}$$

$$=\lim_{R\to\infty}\int_0^{\pi}\frac{P(R\,e^{i\theta})}{Q(R\,e^{i\theta})}iR\,e^{i\theta}d\theta$$

$$=2\pi i\sum_{\mathrm{I}[z]>0}\mathrm{Res}\left[\frac{P(z)}{Q(z)}\right],$$

where $Res[z]$ denotes the complex residues. Solving,

$$\lim_{R\to\infty}\int_{-R}^{R}\frac{P(z)dz}{Q(z)}=2\pi i\sum_{\mathrm{I}[z]>0}\mathrm{Res}\frac{P(z)}{Q(z)}-\lim_{R\to\infty}\int_0^{\pi}\frac{P(R\,e^{i\theta})}{Q(R\,e^{i\theta})}iR\,e^{i\theta}d\theta.$$

Define,

$$I_R=\lim_{R\to\infty}\int_0^{\pi}\frac{P(R\,e^{i\theta})}{Q(R\,e^{i\theta})}iR\,e^{i\theta}d\theta$$

$$=\lim_{R\to\infty}\int_0^{\pi}\frac{b_n(R\,e^{i\theta})^n+b_{n-1}(R\,e^{i\theta})^{n-1}+\ldots+b_0}{c_m(R\,e^{i\theta})^m+c_{m-1}(R\,e^{i\theta})^{m-1}+\ldots+c_0}iR\,d\theta$$

$$\lim_{R\to\infty}\int_0^{\pi}\frac{b_n}{c_m}\left(\mathrm{R}e^{i\theta}\right)^{n-m}iR\,d\theta$$

$$\lim_{R\to\infty}\int_0^{\pi}\frac{b_n}{c_m}R^{n+1-m}\,i(e^{i\theta})^{n-m}\,d\theta$$

and set,

$$\in\ \equiv-(n+1-m),$$

then equation becomes,

$$I_R=\lim_{R\to\infty}\frac{i}{R^{\in}}\frac{b_n}{c_m}\int_0^{\pi}e^{i(n-m)}d\theta$$

Now,

$$\lim_{R\to\infty}R^{-\in}=0$$

for $\in>0$ That means that for $-n-1+m\geq 1$, or $m\geq n+2$, I_R, so

$$\int_{-\infty}^{\infty}\frac{P(z)dz}{Q(z)}=2\pi i\sum_{\mathrm{I}[z]>0}\mathrm{Res}\left[\frac{P(z)}{Q(z)}\right]$$

For $m \geq n+2$. Apply Jordan's lemma with $f(x) \equiv P(x)/Q(x)$. We must have

$$\lim_{x\to\infty} f(x) = 0,$$

so we require $m \geq n+1$.

Then,

$$\int_{-\infty}^{\infty} \frac{P(z)dz}{Q(z)} e^{iaz} = 2\pi i \sum_{\mathrm{I}[z]>0} \mathrm{Res}\left[\frac{P(z)}{Q(z)} e^{iaz}\right]$$

for $m \geq n+1$ and $a > 0$. Since this must hold separately for real and imaginary parts, this result can be extended to,

$$\int_{-\infty}^{\infty} \frac{P(z)}{Q(z)} \cos(ax)\, dx = 2\pi \mathrm{R}\left\{\sum_{\mathrm{I}[z]>0} \mathrm{Res}\left[\frac{P(z)}{Q(z)} e^{iaz}\right]\right\}$$

$$\int_{-\infty}^{\infty} \frac{P(z)}{Q(z)} \sin(ax)\, dx = 2\pi \mathrm{I}\left\{\sum_{\mathrm{I}[z]>0} \mathrm{Res}\left[\frac{P(z)}{Q(z)} e^{iaz}\right]\right\}.$$

Vector Calculus

Vector calculus is concerned with differentiation and integration of vector fields, primarily in 3-dimensional Euclidean space The term "vector calculus" is sometimes used as a synonym for the broader subject of multivariable calculus.

Scalars are quantities that only have a magnitude like mass, speed, and electric field strength. Many times it is often useful to have a quantity that has not only a magnitude but also a direction; such a quantity is called a vector. Examples of quantities represented by vectors include velocity, acceleration, and virtually any type of force (frictional, gravitational, electric, magnetic, etc.). Note that all of these quantities not only have a magnitude (such as speed - the madnitude of the velocity vector) but also occur or act in a given direction.

As an example of when vectors are necessary, suppose a plane traveling at 300 mph to the north with no wind present encounters a westerly crosswind of 50 mph. The resultant velocity of the plane is the sum of the velocities of the wind and the plane. To find this resultant velocity, we must use vectors.

We represent vectors with ordered pairs in pointed brackets to distinguish them from ordered pairs in normal parentheses which represent points. The vector <1,4> is a two-dimensional vector, or directed line segment, from any point (x,y) to the point (x+1,y+4). Likewise, the vector <a,b,c> is a three-dimensional vector from any point (x,y,z) to the point (x+a,y+b,z+c). It is important to remember that a vector is independent of its position in the coordinate system.

The magnitude (or length) of a vector v with initial point (x_1,y_1,z_1) and terminal point (x_2,y_2,z_2) is,

$$\left|\vec{v}\right| = \sqrt{(x_2 - x_1)^2 + (y_2 - y_1)^2 + (z_2 - z_1)^2}$$

Vectors obey the natural intuitive laws of addition and scalar multiplication:

$$\varepsilon < x_1, y_1 > + d < x_2, y_2 > = < \varepsilon x_1, d x_2, \varepsilon y_1, dy_2 >$$

The figures below illustrate the operations of addition and scalar multiplication in the two-dimensional case.

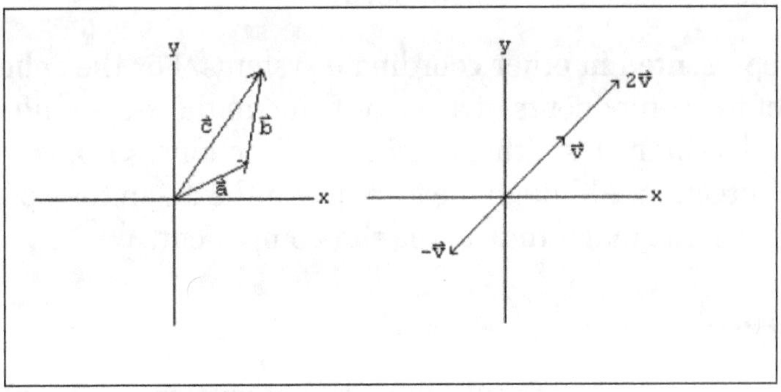

Addition of vectors Scalar Multiplication

The vectors i=<1,0,0>, j=<0,1,0>, and k=<0,0,1> are special since the have unit length and point in the directions of the x-, y-, and z-axes. Any vector in three dimensions can be represented as a linear combination of these three vectors:

$$< x, y, z > = x < 1,0,0 > + y < 0,1,0 > + z < 0,0,1 >$$

The x-, y-, and z-components of a vector are the vectors x<1,0,0>, y<0,1,0>, and z<0,0,1>, respectively.

Example

The vector from the origin to a point P=(x,y,z) has a special name. It is called the position vector of the point since it describes the position of the point relative to the origin.

Example

To solve the example given above involving the plane, we define the direction <1,0> to be east and the direction <0,1> to be north and thus represent the velocity of the plane by p=<300,0> and the velocity of the wind by w=<0,-50>. The resultant velocity of the plane is the sum of these two vectors: r=<300,0>+<0,-50>=<300,-50>. Thus, the plane actually travels in a direction to the west of north.

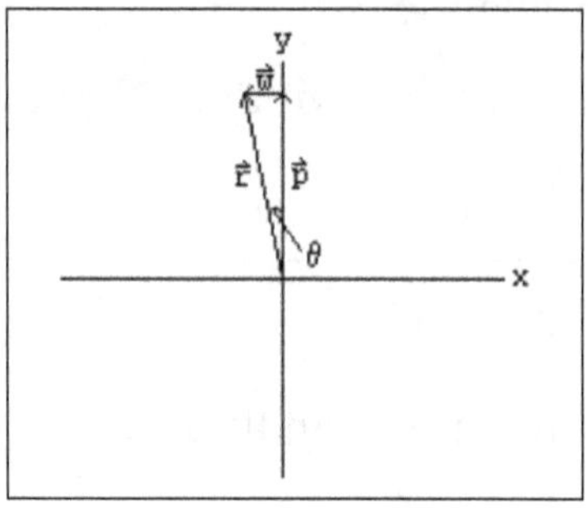

The vertical component of this resultant vector is 300<0,1> and the horizontal component is -50<1,0>. Using the figure above we see that tan(θ)=-50/300=-1/6 which implies theta=-9.5 degrees. Thus, using the diagram as a guide, we see that the resulting motion of the plane is 9.5 degrees west of north.

Vectors can be represented in other coordinate systems. For the spherical coordinate system, instead of the components of a vector being in the x-, y-, and z-directions, the components would be in the rho-, theta-, and phi-directions. <1,0,0> would be the unit vector in the rho-direction. Although the formula for the magnitude of a vector is much simpler since it is just the magnitude of the rho-component.

Vector Functions

A vector function covers a set of multidimensional vectors at the intersection of the domains of $f, g, and\ h.$

Vector valued functions, also called vector functions, allow you to express the position of a point in multiple dimensions within a single function. These can be expressed in an infinite number of dimensions, but are most often expressed in two or three. The input into a vector valued function can be a vector or a scalar. In this atom we are going to introduce the properties and uses of the vector valued functions.

Properties of Vector Valued Functions

A vector valued function allows you to represent the position of a particle in one or more dimensions. A three-dimensional vector valued function requires three functions, one for each dimension. In Cartesian form with standard unit vectors (i,j,k), a vector valued function can be represented in either of the following ways:

$$\mathrm{r}(t) = f(t)\mathrm{i} + g(t)\mathrm{j} + h(t)\mathrm{k}$$

$$\mathrm{r}(t) = \langle f(t), g(t), h(t) \rangle$$

where t is being used as the variable. This is a three dimensional vector valued function. The graph shows a visual representation of-

$$r(t) = \langle 2\cos(t), 4\sin(t), t\rangle$$

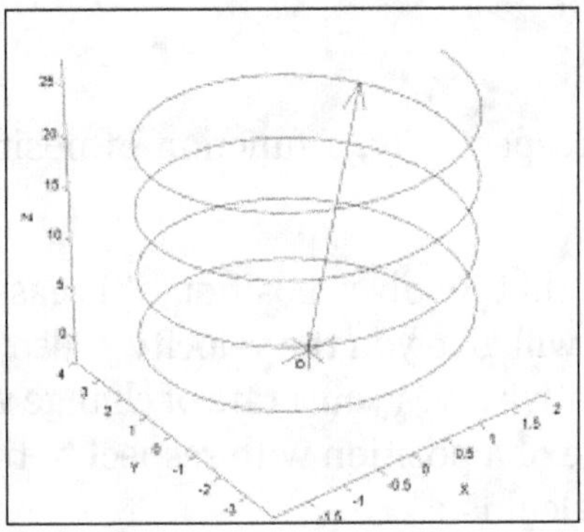

Vector-Valued Function: This a graph of a parametric curve (a simple vector-valued function with a single parameter of dimension 1.) The graph is of the curve: $\langle 2\cos(t), 4\sin(t), t\rangle$ where t goes from 0 to 8π.

This can be broken down into three separate functions called component functions:

$x(t) = 2\cos(t)\, y(t) = 4\sin(t)\, z(t) = t.$

If you were to take a cross section, with the cut perpendicular to any of the three axes, you would see the graph of that function. For example, if you were to slice the three-dimensional shape perpendicular to the z -axis,the graph you would see would be of the function $z(t) = t$. The domain of a vector valued function is a domain that satisfies all of the component functions. It can be found by taking the intersection of the individual component function domains. The vector valued functions can be manipulated in the same way as a vector; they can be added, subtracted, and the dot product and the cross product can be found.

Example

For this example, we will use time as our parameter. The following vector valued function represents time, t:

$$r(t) = f(t)i + g(t)j + h(t)k$$

This function is representing a position. Therefore, if we take the derivative of this function, we will get the velocity:

$$\frac{dr(t)}{dt} = f(t)i' + g(t)j' + h(t)k'$$
$$= v(t)$$

If we differentiate a second time, we will be left with acceleration:

$$\frac{dv(t)}{dt} = a(t)$$

Arc Length and Speed

Arc length and speed are, respectively, a function of position and its derivative with respect to time.

Since length is a magnitude that involves position, it is easy to deduce that the derivative of a length, or position, will give you the velocity —also known as speed—of a function. This is because a derivative gives you a rate of change with respect to a parameter. Velocity is the rate of change of a position with respect to time. Let's start this atom by looking at arc length with calculus.

Arc Length

The arc length is the length you would get if you took a curve, straightened it out, and then measured the length of that line. The arc length can be found using geometry, but for the sake of this atom, we are going to use integration. The arc length is approximated by connecting a finite number of points along and curve, connecting those lines to create a a string of very small straight lines, and adding them together. To find this using integration, we should start out by using the Pythagorean Theorem for length of the different sides of a triangle:

$$ds^2 = dx^2 + dy^2 \frac{ds^2}{dx^2}$$

$$= 1 + \frac{dy^2}{dx^2} ds$$

$$= \sqrt{1 + \frac{dy^{2^2}}{dx^2}} \cdot dx$$

$$= \int_a^b \sqrt{1 + f^1(x)^2} \cdot dx$$

where is the arc length. If $x \quad X(t)$ and $y = Y(t)$,

$$s = \int_a^b \sqrt{1 + f'(x)^2} \cdots dx$$

$$= \int_a^b \sqrt{\left[X^1(t)^2\right] + \left[Y^1(t)\right]} \cdots dt$$

$$= \int_a^b \sqrt{dx^2 + dy^2}$$

$$= \int_a^b \sqrt{\frac{dx^2}{dt} + \frac{dy^2}{dt} \cdots dt}$$

Since this is a function of position and is defined by x, we need to have a derivative that is in respect to x:

$$\int_a^b \sqrt{1 + \frac{dy^2}{dx} * dx}$$

$$r = f(\theta)$$

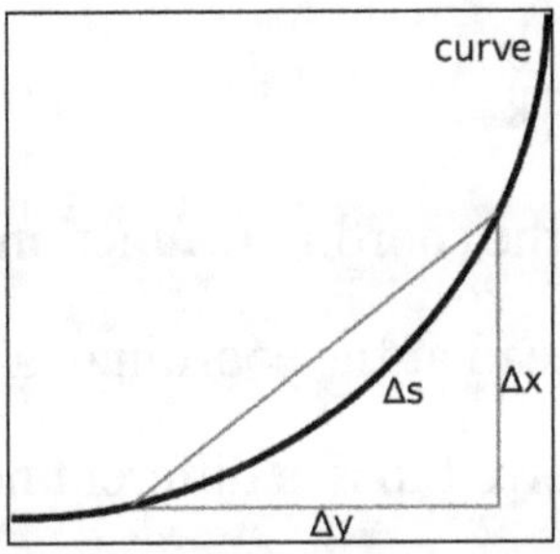

Curves and the Pythagorean Theorem: For a small piece of curve, Δs can be approximated with the Pythagorean theorem.

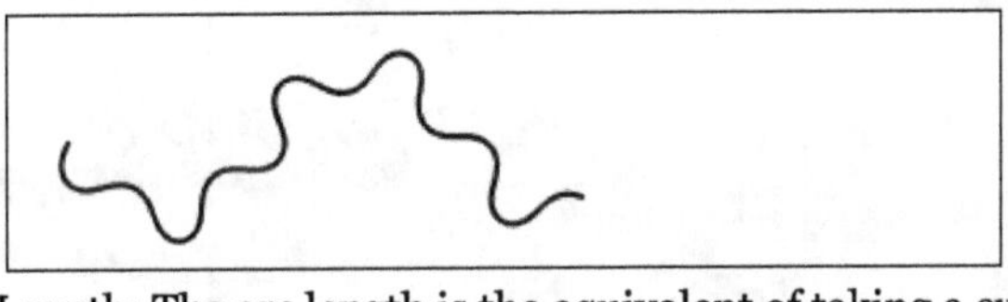

Arc Length: The arc length is the equivalent of taking a curve, straightening it out, and then measuring it.

Arc Speed

Now that the hard part is over, we can easily find the speed along this curve. Since speed is in relation to time and not position, we need to revert back to the arc length with respect to time:

$$\int_a^b \sqrt{\frac{dx^2}{dt} + \frac{dy^2}{dt} \cdot dt}$$

Then, differentiate with respect to time:

$$v(t) = s' \sqrt{\left[X^1(t)^2\right] + \left[Y^1(t)\right]^2}$$

Calculus of Vector-Valued Functions

A vector function is a function that can behave as a group of individual vectors and can perform differential and integral operations.

A vector-valued function, also referred to as a vector function, is a mathematical function of one or more variables whose range is a set of multidimensional vectors or infinite-dimensional vectors. The input of a vector-valued function could be a scalar or a vector. The dimension of the domain is not defined by the dimension of the range.

A common example of a vector valued function is one that depends on a single real number parameter t, often representing time, producing a vector $v(t)$ as the result. In terms of the standard unit vectors i, j, k of Cartesian 3-space, these specific type of vector-valued functions are given by expressions such as:

$$\mathrm{r}(t) = f(t)\mathrm{i}, \mathrm{g}(t)\mathrm{j} + h(t)\mathrm{k}$$

where $f(t), g(t)$, and $h(t)$ are the coordinate functions of the parameter t. The vector has its tail at the origin and its head at the coordinates evaluated by the function.

Vector functions can also be referred to in a different notation:

$$\mathrm{r}(t) = \langle f(t), \mathrm{g}(t), h(t) \rangle$$

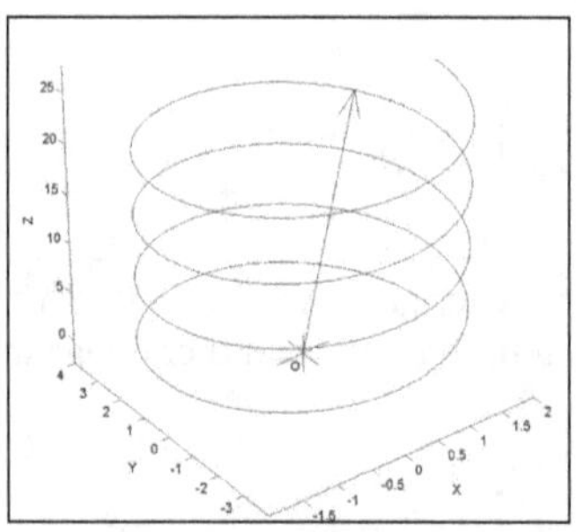

Vector valued function: This graph is a visual representation of the three-dimensional vector-valued function $r(t) = \langle 2\cos(t), 4\sin(t), t \rangle$. This can be broken down into three separate functions called component functions: $x(t) = 2\cos(t)\, y(t) = 4\sin(t)\, z(t) = t$.

Vector calculus is a branch of mathematics that covers differentiation and integration of vector fields in any number of dimensions. Because vector functions behave like individual vectors, you can manipulate them the same way you can a vector. Vector calculus is used extensively throughout physics and engineering, mostly with regard to electromagnetic fields, gravitational fields, and fluid flow. When taking the derivative of a vector function, the function should be treated as a group of individual functions.

Vector functions are used in a number of differential operations, such as gradient (measures the rate and direction of change in a scalar field), curl (measures the tendency of the vector function to rotate about a point in a vector field), and divergence (measures the magnitude of a source at a given point in a vector field).

Arc Length and Curvature

The curvature of an object is the degree to which it deviates from being flat and can be found using arc length.

Arc Curvature

The curvature of an arc is a value that represents the direction and sharpness of a curve. On any curve, there is a center of curvature, C. This is the intersection point of two infinitely close normals to this curve. The radius, R, is the distance from this intersection point to the center of curvature.

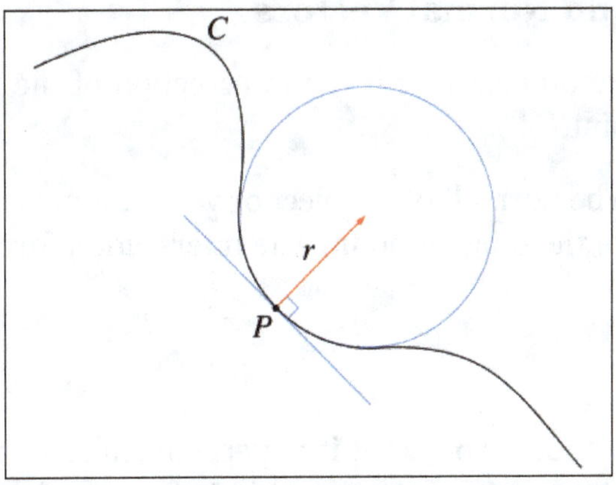

Curvature: Curvature is the amount an object deviates from being flat. Given any curve C and a point P on it, there is a unique circle or line which most closely approximates the curve near P\. The curvature of C at P is then defined to be the curvature of that circle or line. The radius of curvature is defined as the reciprocal of the curvature.

In order to find the value of the curvature, we need to take the parameter time, s, and the unit tangent vector, which in this case is the same as the unit velocity vector, T, which is also a function of time.The curvature is a magnitude of the rate of change of the tangent vector, T:

$$k = \|\frac{dT}{ds}\|$$

Where κ is the curvature and $\frac{dT}{ds}$ is the acceleration vector (the rate of change of the velocity vector over time).

relation between Curvature and Arc Length

The curvature can also be approximated using limits. Given the points P and Q on the curve, lets call the arc length s(P,Q), and the linear distance from P to Q will be denoted as d(P,Q). The curvature of the arc at point P can be found by obtaining the limit:

$$\kappa(P) = \frac{\lim}{Q \to P}\sqrt{\frac{24*(s(P,Q) - d(P,Q))}{s(P,Q)^3}}$$

In order to use this formula, you must first obtain the arc length of the curve from points P to Q and length of the linear segment that connect points P and Q.In Cartesian coordinates:

$$\int_a^b \sqrt{1 + \frac{dy^2}{dx}} * dx$$

Tangent Vectors and Normal Vectors

A vector is normal to another vector if the intersection of the two form a 90-degree angle at the tangent point.

In order for a vector to be normal to an object or vector, it must be perpendicular with the directional vector of the tangent point. The intersection formed by the two objects must be a right angle.

Normal Vectors

An object is normal to another object if it is perpendicular to the point of reference. That means that the intersection of the two objects forms a right angle. Usually, these vectors are denoted as n.

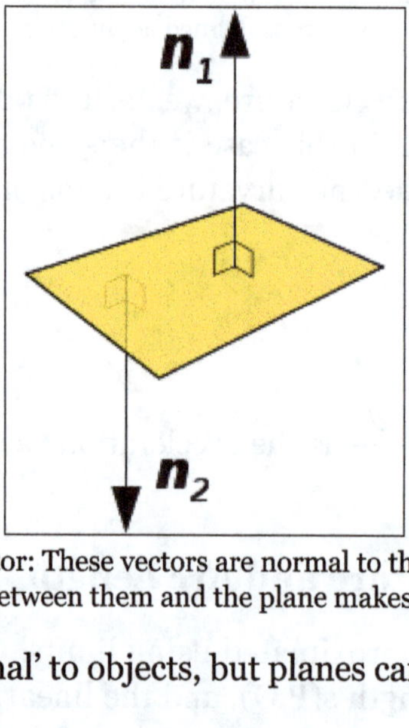

Figure: Normal Vector: These vectors are normal to the plane because the intersection between them and the plane makes a right angle.

Not only can vectors be 'normal' to objects, but planes can also be normal.

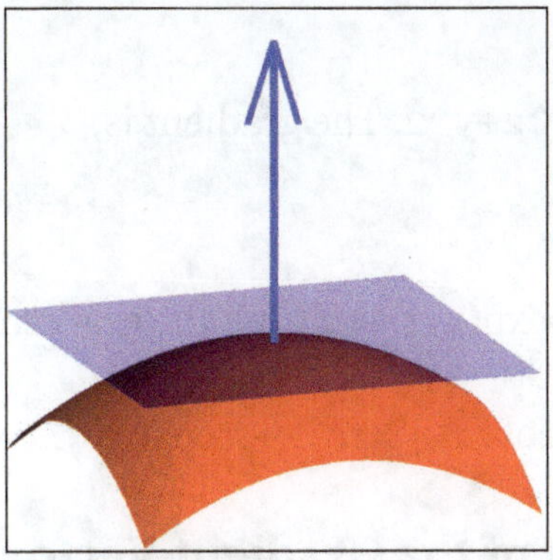

Figure: Normal Plane: A plane can be determined as normal to the object if the directional vector of the plane makes a right angle with the object at its tangent point. This plane is normal to the point on the sphere to which it is tangent. Each point on the sphere will have a unique normal plane.

Dot Product

As we covered in another atom, one of the manipulations of vectors is called the Dot Product. When you take the dot product of two vectors, your answer is in the form of a single value, not a vector. In order for two vectors to be normal to each other, the dot product has to be zero.

$$\begin{aligned} a \bullet b &= 0 \\ &= a_1b_1 + a_2b_2 + a_3b_3 \\ &= |a||b|\cos\theta \end{aligned}$$

Tangent Vectors

Tangent vectors are almost exactly like normal vectors, except they are tangent instead of normal to the other vector or object. These vectors can be found by obtaining the derivative of the reference vector, $r(t)$

$$r(t) = f(t)i + g(t)j + h(t)k$$

Gradient and Directional Derivative

The gradient of a function w=f(x,y,z) is the vector function:

$$\nabla f = grad\ f = < \frac{\partial f}{\partial x}(x,y,z), \frac{\partial f}{\partial y}(x,y,z), \frac{\partial f}{\partial z}(x,y,z) >$$

For a function of two variables z=f(x,y), the gradient is the two-dimensional vector <f_x(x,y),f_y(x,y)>. This definition generalizes in a natural way to functions of more than three variables.

Examples

For the function z=f(x,y)=4x^2+y^2. The gradient is,

$$grad\ f = < 8x, 2y >.$$

For the function w=g(x,y,z)=exp(xyz)+sin(xy), the gradient is

$$grad\ g = < yze^{xyz} + y\cos(xy), xze^{xyz} + x\cos(xy), xye^{xyz} >.$$

Geometric Description of the Gradient Vector

There is a nice way to describe the gradient geometrically. Consider $z = f(x, y) = 4x^2 + y^2$. The surface defined by this function is an elliptical paraboloid. This is a bowl-shaped surface. The bottom of the bowl lies at the origin. The figure below shows the level curves, defined by f(x,y)=c, of the surface. The level curves are the ellipses $4x^2 + y^2 = c$.

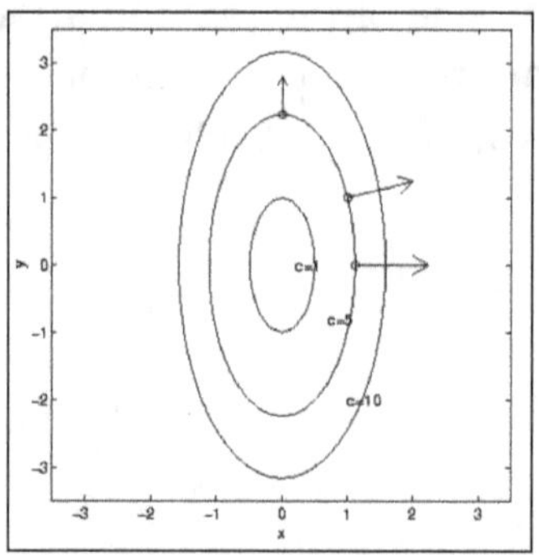

The gradient vector <8x,2y> is plotted at the 3 points (sqrt(1.25),0), (1,1), (0,sqrt(5)). As the plot shows, *the gradient vector at (x,y) is normal to the level curve through (x,y)*. As we will see below, the gradient vector points in the direction of greatest rate of increase of f(x,y)

In three dimensions the level curves are level surfaces. Again, the gradient vector at (x,y,z) is normal to level surface through (x,y,z).

Directional Derivatives

For a function z=f(x,y), the partial derivative with respect to x gives the rate of change of f in the x direction and the partial derivative with respect to y gives the rate of change of f in the y direction. How do we compute the rate of change of f in an arbitrary direction?

The rate of change of a function of several variables in the direction u is called the *directional derivative* in the direction u. Here u is assumed to be a unit vector. Assuming w=f(x,y,z) and u=<u_1,u_2,u_3>, we have

$$D_u f = \text{grad } f.u = \frac{\partial f}{\partial x}u_1 + \frac{\partial f}{\partial y}u_2 + \frac{\partial f}{\partial y}u_3.$$

Hence, the directional derivative is the dot product of the gradient and the vector u. Note that if u is a unit vector in the x direction, u=<1,0,0>, then the directional derivative is simply the partial derivative with respect to x. For a general direction, the directional derivative is a combination of the all three partial derivatives.

Example

What is the directional derivative in the direction <1,2> of the function z=f(x,y)=4x-^2+y^2 at the point x=1 and y=1. The gradient is <8x,2y>, which is <8,2> at the point x=1 and y=1. The direction u is <2,1>. Converting this to a unit vector, we have <2,1>/ sqrt(5). Hence,

$$D_u f = \text{grad } f \cdot u = <8,2>.<\frac{2}{\sqrt{5}}, \frac{1}{\sqrt{5}}> = \frac{18}{\sqrt{5}}.$$

Directions of Greatest Increase and Decrease

The directional derivative can also be written:

$$D_u f = \text{grad } f \cdot u = |\text{grad } f||u|\cos\theta$$

where theta is the angle between the gradient vector and u. The directional derivative takes on its greatest positive value if theta=0. Hence, the direction of greatest increase of f is the same direction as the gradient vector. The directional derivative takes on its greatest negative value if theta=pi (or 180 degrees). Hence, the direction of greatest decrease of f is the direction opposite to the gradient vector.

Curl

The curl of a vector field, denoted $\text{curl}(\mathrm{F})$ or $\nabla \times \mathrm{F}$ (the notation used in this work), is defined as the vector field having magnitude equal to the maximum “circulation” at each point and to be oriented perpendicularly to this plane of circulation for each point. More precisely, the magnitude of $\nabla \times \mathrm{F}$ is the limiting value of circulation per unit area. Written explicitly,

$$(\nabla \times \mathrm{F}).\hat{\mathrm{n}} \equiv \lim_{A \to 0} \frac{\oint_C \mathrm{F} \cdot d\mathrm{s}}{A}.$$

where the right side is a line integral around an infinitesimal region of area A that is allowed to shrink to zero via a limiting process and $\hat{\mathrm{n}}$ is the unit normal vector to this region. If $\nabla \times \mathrm{F} = 0$, then the field is said to be an irrotational field. The symbol ∇ is variously known as "nabla" or "del."

The physical significance of the curl of a vector field is the amount of "rotation" or angular momentum of the contents of given region of space. It arises in fluid mechanics and elasticity theory. It is also fundamental in the theory of electromagnetism, where it arises in two of the four Maxwell equations,

$$\nabla \times \mathrm{E} = -\frac{\partial B}{\partial_t}$$

$$\nabla \times \mathrm{B} = \mu_0 \mathrm{J} + \epsilon_0 \mu_0 \frac{\partial \mathrm{E}}{\partial_t},$$

where MKS units have been used here, E denotes the electric field, B is the magnetic field, μ_0 is a constant of proportionality known as the permeability of free space, J is the current density, and ϵ_0 is another constant of proportionality known as the permittivity of free space. Together with the two other of the Maxwell equations, these formulas describe virtually all classical and relativistic properties of electromagnetism.

In Cartesian coordinates, the curl is defined by,

$$\nabla \times \mathrm{F} = \left(\frac{\partial F_Z}{\partial y} - \frac{\partial F_y}{\partial z}\right)\hat{\mathrm{x}} + \left(\frac{\partial F_x}{\partial z} - \frac{\partial F_z}{\partial x}\right)\hat{\mathrm{y}} + \left(\frac{\partial F_y}{\partial x} - \frac{\partial F_x}{\partial y}\right)\hat{z}.$$

This provides the motivation behind the adoption of the symbol $\nabla \times$ for the curl, since interpreting ∇ as the gradient operator $\nabla = (\partial/\partial x, \partial/\partial y, \partial/\partial z)$, the "cross product" of the gradient operator with F is given by,

$$\nabla \times \mathrm{F} = \begin{vmatrix} \hat{\mathrm{x}} & \hat{\mathrm{y}} & \hat{z} \\ \frac{\partial}{\partial x} & \frac{\partial}{\partial y} & \frac{\partial}{\partial z} \\ F_x & F_y & F_z \end{vmatrix},$$

which is precisely equation (4). A somewhat more elegant formulation of the curl is given by the matrix operator equation,

$$\nabla \times \mathrm{F} = \begin{vmatrix} 0 & -\frac{\partial}{\partial z} & \frac{\partial}{\partial y} \\ \frac{\partial}{\partial z} & 0 & -\frac{\partial}{\partial x} \\ -\frac{\partial}{\partial y} & \frac{\partial}{\partial x} & 0 \end{vmatrix} \mathrm{F}$$

The curl can be similarly defined in arbitrary orthogonal curvilinear coordinates using

$$\mathrm{F} \equiv F_1 \hat{\mathrm{u}}_1 + F_2 \hat{\mathrm{u}}_2 + F_3 \hat{\mathrm{u}}_3$$

and defining,

$$h_i \equiv \left| \frac{\partial \mathrm{r}}{\partial u_i} \right|,$$

As,

$$\nabla \times \mathrm{F} =\} \frac{1}{h_1 h_2 h_3} \begin{vmatrix} h_1 \hat{\mathrm{u}}_1 & h_2 \hat{\mathrm{u}}_2 & h_3 \hat{\mathrm{u}}_3 \\ \frac{\partial}{\partial u_1} & \frac{\partial}{\partial u_2} & \frac{\partial}{\partial u_3} \\ h_1 F_1 & h_2 F_2 & h_3 F_3 \end{vmatrix}$$

$$= \frac{1}{h_2 h_3} \left[\frac{\partial}{\partial u_2}(h_3 F_3) - \frac{\partial}{\partial u_3}(h_2 F_2) \right] \hat{u}_1 + \frac{1}{h_1 h_3}$$

$$\left[\frac{\partial}{\partial u_3}(h_1 F_1) - \frac{\partial}{\partial u_1}(h_3 F_3) \right] \hat{u}_2 + \frac{1}{h_1 h_2} \left[\frac{\partial}{\partial u_1}(h_2 F_2) - \frac{\partial}{\partial u_2}(h_1 F_1) \right] \hat{u}_3.$$

The curl can be generalized from a vector field to a tensor field as,

$$(\nabla \times A)^a = \in^{a\mu\nu} A_{y:\mu},$$

where $\in_{ijk}$ is the permutation tensor and ";" denotes a covariant derivative.

Fundamental Theorem of Calculus

The Fundamental Theorem of Calculus justifies the procedure of evaluating an antiderivative at the upper and lower limits of integration and taking the difference.

The First Fundamental Theorem of Calculus

We have learned about indefinite integrals, which was the process of finding the antiderivative of a function. In contrast to the indefinite integral, the result of a definite integral will be a number, instead of a function. The definite integral of a function is the signed area under the graph of the function, and is expressed in the form of $\int_a^b f(x)\,dx$:

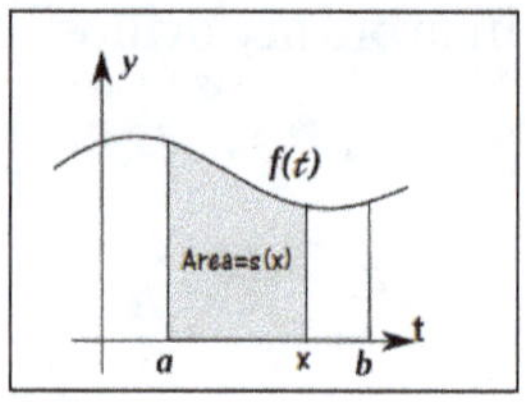

Now, suppose that we formed an area function $S(x)$ in such a way that it is dependent on the function $f(x)$ as,

$$S(x)=\int_a^x f(x)dt,$$

where f is continuous on the interval $[a,b]$. Now, suppose we wanted to find the the rate of change of the area with respect to x,

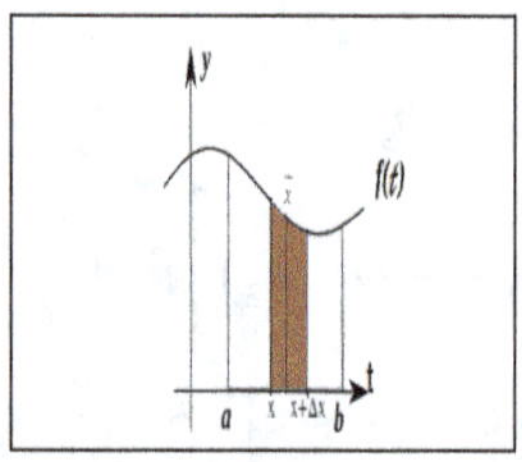

We can see from the figure above that the area of the shaded region is equal to the area under the curve $f(t)$ from a $x+\Delta x$ minus the area under $f(t)$ from a to x. Thus,

$$\Delta S = A(x+\Delta x)-A(x)$$

$$\frac{\Delta S}{\Delta x}=\frac{A(x+\Delta x)-A(x)}{\Delta x}.$$

So, the rate of change of area becomes,

$$S'(x)=\frac{ds}{dx}=\lim_{\Delta x\to 0}\frac{S(x+\Delta x)-S(x)}{\Delta x}.$$

We know that there is an $\overline{x}$ found between x and $x+\Delta x$ such that the area of the shaded region is equal $f(\overline{x})\Delta x$:

$$\begin{aligned}
S'(x) &= \lim_{\Delta x\to 0}\frac{S(x+\Delta x)-S(x)}{\Delta x} \\
&= \lim_{\Delta x\to 0}\frac{f(\overline{x})\Delta x}{\Delta x} \\
&= \lim_{\Delta x\to 0} f(\overline{x}) \\
&= f(x).
\end{aligned}$$

The last step is true because, as $\Delta x \to 0$, anything found between x and $x+\Delta x$ approaches x. So, now we are ready to state the first fundamental theorem of calculus.

If f is continuous on $[a,b]$, then the function defined by,

$$S(x)=\int_a^x f(t)\,dt$$

is continuous on$[a,b]$ and differentiable on $a,b)$, and $S'(x)=f(x)$.

So basically integration is the opposite of differentiation. More clearly, the first fundamental theorem of calculus can be rewritten in Leibniz notation as,

$$\frac{d}{dx}\int_a^x f(t)\,dt=f(x).$$

Find the derivative of $k(x)=\int_2^x\left(4^t+t\right)dt.$

The function f is continuous, so from the first fundamental theorem of calculus we have,

$$k'(x)=4^x+x.$$

What is the derivative of $h(x)=\int_2^{x^2}\frac{1}{1+t^2}dt$?

We use the first fundamental theorem of calculus in accordance with the chain-rule to solve this.

Let $u=x^2$, then

$$\frac{d}{dx}\int_2^{x^2}\frac{1}{1+t^2}dt=\frac{d}{du}\left[\int_1^u\frac{1}{1+t^2}dt\right],\frac{du}{dx}$$

$$=\frac{1}{1+u^2}.2x$$

$$=\frac{2x}{1+x^4}.$$

Find the derivative of $\int_2^{x^2}\cos t\,dt$.

Again, we use the chain rule along with the fundamental theorem of calculus to solve this.

Let $u=x^2$, then

$$\int_1^{x^2} \cos t \, dt = \frac{d}{du}\left[\int_1^u \cos t \, dt\right] . \frac{du}{dx}$$
$$= \cos u . \frac{d}{dx}\left(x^2\right)$$
$$= \cos u . 2x$$
$$= 2x \cos x^2.$$

Find the derivative of $h(x) = \int_x^{3x} \sin\theta \, d\theta$

We use the following property of integrals:

$$\int_b^a f(x)\, dx = \int_b^0 f(x)\, dx + \int_0^a f(x)\, dx.$$

So,

$$h'(x) = \frac{d}{dx}\int_x^{3x} \sin\theta \, d\theta$$
$$= \frac{d}{dx}\int_x^0 \sin\theta \, d\theta + \frac{d}{dx}\int_0^{3x} \sin\theta \, d\theta$$
$$= \frac{d}{dx} - \int_0^x \sin\theta \, d\theta + \frac{d}{dx}\int_0^{3x} \sin\theta \, d\theta$$
$$= -\sin x + \frac{d}{du}\int_0^u \sin\theta \, d\theta . \frac{du}{dx}$$
$$= -\sin x + 3\sin 3x.$$

Second Fundamental Theorem of Calculus

If f is a continuous function on$[a,b]$, then

$$\int_a^b f(x) dx = F(b) - F(a),$$

where F is an anti-derivative of f, i.e. $F'=f$.

We know from the first fundamental theorem of calculus that if $s(x) = \int_a^x f(x) dt$ then S is an anti- derivative of f or S′(x)=f(x). And since,F′(x)=f(x),

$$S'(x) = F'(x).$$

Integrating both sides with respect to x, we have

$$S(x)=\int_a^x f(t)dt=F(x)+C,$$

where C is some constant.

Now, plugging in $x=a$ in this equation, we have

$$S(a)=\int_a^a f(t)dt=F(a)+C.$$

By definition, $S(a)=\int_a^a f(t)dt=0$, and hence

$$F(a)+C=0$$
$$\Rightarrow C=-F(a).$$

Plugging in $C=-F(a)$ back in the equation($S(x)=\int_a^x f(t)dt=F(x)+C,$), we have

$$S(x)=\int_a^x f(t)dt=F(x)-F(a).$$

Finally, setting $x=b$, we have

$$S(b)=\int_a^b f(t)dt=F(b)-F(a)$$
$$\Rightarrow \int_a^b f(x)dx=F(b)-F(a).$$

We could also prove the above theorem using the concept of Riemann sums.

We know from the first fundamental theorem of calculus that if $S(x)=\int_a^x f(t)dt$, then S is an anti-derivative of f or $S'(x)=f(x)$ So we know that $f(x)=F'(x)$. Consider a partition $P=\{a=x_0<x_1<x_2<\cdots<xn=b\}$ With this partition as reference, we can write,

$$F(b)-F(a)=\sum_{i=0}^{n-1}\left[F(x_{i+1})-F(x_i)\right].$$

This is now the neater part of the proof. $\forall i=0,1,2,\ldots,n-1\ \exists t_i\in(x_i,x_{i+1})$ such that $F(x_{i+1})-F(x_i)=F'(t_i)(x_{i+1}-x_i)=f(t_i)(x_{i+1}-x_i)$. This is direct implication of the mean value theorem. So now what we have is,

$$F(b)-F(a)=\sum_{i=0}^{n-1} f(t_i)(x_{i+1}-x_i);t_i\in(x_i,x_{i+1}).$$

The right-hand side of the expression is nothing but the Riemann sum which will eventually converge to definite integral as the partition P gets finer and finer:

$$F(b)-F(a)=\int_a^b f(t)dt.$$

This theorem transforms the difficult problem of evaluating definite integrals by calculating limits of sums, into an easier problem of finding an anti-derivative. So for example if we are asked to compute the integral $\int_a^b f(x)\ dx$, we find an anti-derivative of $f(x)$ and compute their value at each end-point of the integral, and finally subtract them from each other.

References

- Definition-of-calculus: thoughtco.com, Retrieved 14 May, 2019
- Differential_calculus: encyclopediaofmath.org, Retrieved 5 February, 2019
- Limits, limits-and-derivatives, maths: toppr.com, Retrieved 26 August, 2019
- Integral_calculus: encyclopediaofmath.org, Retrieved 22 April, 2019
- RiemannIntegral: wolfram.com, Retrieved 28 June, 2019
- Lebesgue-integral, science: britannica.com, Retrieved 23 May, 2019
- ContourIntegration: wolfram.com, Retrieved 20 June, 2019
- Vector, calculus: oregonstate.edu, Retrieved 10 January, 2019
- Vector-functions, boundless-calculus: lumenlearning.com, Retrieved 19 July, 2019
- Curl: wolfram.com, Retrieved 25 March, 2019
- Fundamental-theorem-of-calculus: brilliant.org, Retrieved 6 April, 2019

6

Differential Equations

A mathematical equation that relates some function with its derivatives is known as a differential equation. Ordinary differential equations, partial differential equation and non-linear differential equation are the common types of differential equations. The diverse applications of differential equation in the current scenario have been thoroughly discussed in this chapter.

Differential equation is mathematical statement containing one or more derivatives—that is, terms representing the rates of change of continuously varying quantities. Differential equations are very common in science and engineering, as well as in many other fields of quantitative study, because what can be directly observed and measured for systems undergoing changes are their rates of change. The solution of a differential equation is, in general, an equation expressing the functional dependence of one variable upon one or more others; it ordinarily contains constant terms that are not present in the original differential equation. Another way of saying this is that the solution of a differential equation produces a function that can be used to predict the behaviour of the original system, at least within certain constraints.

Differential equations are classified into several broad categories, and these are in turn further divided into many subcategories. The most important categories are ordinary differential equations and partial differential equations. When the function involved in the equation depends on only a single variable, its derivatives are ordinary derivatives and the differential equation is classed as an ordinary differential equation. On the other hand, if the function depends on several independent variables, so that its derivatives are partial derivatives, the differential equation is classed as a partial differential equation. The following are examples of ordinary differential equations:

$$\frac{dy}{dt} = -ky$$

$$m\frac{d^2y}{dt^2} = -k^2y,$$

$$\left[1+\left(\frac{dy}{dx}\right)^2\right]\frac{d^3y}{dx^3} - 3\frac{dy}{dx}\left(\frac{d^2y}{dx^2}\right) = 0.$$

In these, y stands for the function, and either t or x is the independent variable. The symbols k and m are used here to stand for specific constants.

Whichever the type may be, a differential equation is said to be of the nth order if it involves a derivative of the nth order but no derivative of an order higher than this. The equation,

$$\frac{\partial u}{\partial t}=k^2\left[\frac{\partial^2 u}{\partial x^2}+\frac{\partial^2 u}{\partial y^2}+\frac{\partial^2 u}{\partial z^2}\right]$$

is an example of a partial differential equation of the second order. The theories of ordinary and partial differential equations are markedly different, and for this reason the two categories are treated separately.

Instead of a single differential equation, the object of study may be a simultaneous system of such equations. The formulation of the laws of dynamics frequently leads to such systems. In many cases, a single differential equation of the nth order is advantageously replaceable by a system of n simultaneous equations, each of which is of the first order, so that techniques from linear algebra can be applied.

An ordinary differential equation in which, for example, the function and the independent variable are denoted by y and x is in effect an implicit summary of the essential characteristics of y as a function of x. These characteristics would presumably be more accessible to analysis if an explicit formula for y could be produced. Such a formula, or at least an equation in x and y (involving no derivatives) that is deducible from the differential equation, is called a solution of the differential equation.

Ordinary Differential Equation

An ordinary differential equation (frequently called an "ODE," "diff eq," or "diffy Q") is an equality involving a function and its derivatives. An ODE of order n is an equation of the form,

$$F\left(x,y,y',...y^{(n)}\right)=0,$$

where y is a function of $x, y'=dy/dx$ is the first derivative with respect to x, and $y^{(n)}=d^n y/dx^n$ is the nth derivative with respect to x.

Let y be an unknown function,

$$y:\mathbb{R}\rightarrow\mathbb{R}$$

in x with $y^{(i)}$ the i-th derivative of y, then a function

$$F(x,y,y^{(1)},\ldots,y^{(n-1)})=y^{(n)}$$

is called an ordinary differential equation (ODE) of order (or degree) n. For vector valued functions,

$$y:\mathbb{R}\to\mathbb{R}^m$$

we call F a system of ordinary differential equations of dimension m.

A function y is called a solution of F. A general solution of an nth-order equation is a solution containing n arbitrary variables, corresponding to n constants of integration. A particular solution is derived from the general solution by setting the constants to particular values. A singular solution is a solution that can't be derived from the general solution.

When a differential equation of order n has the form,

$$F\left(x,y,y',y'',\ldots,y^{(n)}\right)=0$$

it is called an implicit differential equation whereas the form,

$$F\left(x,y,y',y'',\ldots,y^{(n-1)}\right)=y^{(n)}$$

is called an explicit differential equation.

A differential equation not depending on x is called autonomous.

A differential equation is said to be linear if F can be written as a linear combination of the derivatives of y,

$$y^{(n)}=\sum_{i=1}^{n-1}a_i(x)y^{(i)}+r(x)$$

with $a_i(x)$ and $r(x)$ continuous functions in x. *If* $r(x)=0$ the we call the linear differential equation homogeneous otherwise we call it inhomogeneous.

Examples

Reduction of dimension

Given an explicit ordinary differential equation of order n and dimension 1,

$$F\left(x,y,y',y'',\ldots,y^{(n-1)}\right)=y^{(n)}$$

we define a new family of unknown functions

$$y_n:\ =y^{(n-1)}.$$

We can then rewrite the original differential equation as a system of differential equations with order 1 and dimension n.

$$y_1' = y_2$$
$$\vdots$$
$$y_n' = F(y_n, \ldots, y_1, x).$$

which can be written concisely in vector notation as

$$\mathbf{y}' = \mathbf{F}(\mathbf{y}, x)$$

with

$$\mathbf{y} := (y, \ldots, y_n).$$

Types of Ordinary Differential Equations

Ordinary differential equations which can be categorised by three factors:

- Linear vs. Non-linear.
- Homogeneous vs. Inhomogenous.
- Constant coefficents versus variable coefficients.

Information below provides methods for the solution of these differing ODEs:

Homogeneous Linear ODEs with Constant Coefficients

The first method of solving linear ordinary differential equations with constant coefficients is due to Euler, who realized that solutions have the form e^{zx}, for possibly-complex values of z. Thus to solve,

$$\frac{d^n y}{dx^n} + A_1 \frac{d^{n-1} y}{dx^{n-1}} + \cdots + A_n y = 0$$

we set $y = e^{zx}$, leading to

$$z^n e^{zx} + A_1 z^{n-1} e^{zx} + \cdots + A_n e^{zx} = 0$$

so dividing by e^{zx} gives the nth-order polynomial

$$F(z) = z^n + A_1 z^{n-1} + \cdots + A_n = 0$$

In short the terms,

$$\frac{d^k y}{dx^k} \qquad (k = 1, 2, \cdots, n).$$

of the original differential equation are replaced by z^k. Solving the polynomial gives n values of $z, z_1, \ldots, z_n$. Plugging those values into $e^{z_i x}$ gives a basis for the solution; any linear combination of these basis functions will satisfy the differential equation.

This equation $F(z) = 0$, is the "characteristic" equation considered later by Monge and Cauchy.

Example

$$y'''' - 2y''' + 2y'' - 2y' + y = 0$$

has the characteristic equation

$$z^4 - 2z^3 + 2z^2 - 2z + 1 = 0.$$

This has zeroes, i, $-i$, and 1 (multiplicity 2). The solution basis is then,

$$e^{ix}, e^{-ix}, e^x, xe^x.$$

This corresponds to the real-valued solution basis,

$$\cos x, \sin x, e^x, xe^x.$$

If z is a (possibly not real) zero of $F(z)$ of multiplicity m and $k \in \{0, 1, \ldots, m-1\}$ then $y = x^k e^{zx}$, is a solution of the ODE. These functions make up a basis of the ODE's solutions.

If the A_i are real then real-valued solutions are preferable. Since the non-real z values will come in conjugate pairs, so will their corresponding ys; replace each pair with their linear combinations Re(y) and Im(y).

A case that involves complex roots can be solved with the aid of Euler's formula.

Example: Given $y'' - 4y' + 5y = 0$.The characteristic equation is $z^2 - 4z + 5 = 0$, which has zeroes $2+i$ and $2-i$. Thus the solution basis $\{y_1, y_2\}$ is $\left\{e^{(2+i)x}, e^{(2-i)x}\right\}$.

Now y is a solution if $y = c_1 y_1 + c_2 y_2$ for $c_1, c_2 \in \mathbb{C}$.

Because the coefficients are real,

- We are likely not interested in the complex solutions
- Our basis elements are mutual conjugates

The linear combinations

$$u_1 = \text{Re}(y_1) = \frac{y_1 + y_2}{2} = e^{2x} \cos(x)$$

and

$$u_2 = \mathrm{Im}(y_1) = \frac{y_1 - y_2}{2i} = e^{2x}\sin(x)$$

will give us a real basis in $\{u_1, u_2\}$.

Inhomogeneous Linear ODEs with Constant Coefficients

Suppose instead we face,

$$\frac{d^n y}{dx^n} + A_1 \frac{d^{n-1} y}{dx^{n-1}} + \cdots + A_n y = f(x).$$

For later convenience, define the characteristic polynomial,

$$P(v) = v^n + A_1 v^{n-1} + \cdots + A_n.$$

We find the solution basis $\{y_1, y_2, \ldots, y_n\}$ as in the homogeneous (f=0) case. We now seek a particular solution y_p by the variation of parameters method. Let the coefficients of the linear combination be functions of x:

$$y_p = u_1 y_1 + u_2 y_2 + \cdots + u_n y_n.$$

Using the "operator" notation $D = d/dx$ and a broad-minded use of notation, the ODE in question is $P(D)y = f$;

so,

$$f = P(D)y_p = P(D)(u_1 y_1) + P(D)(u_2 y_2) + \cdots + P(D)(u_n y_n).$$

With the constraints

$$0 = u_1' y_1 + u_2' y_2 + \cdots + u_n' y_n$$
$$0 = u_1' y_1' + u_2' y_2' + \cdots + u_n' y_n'$$

$$0 = u_1' y_1^{(n-2)} + u_2' y_2^{(n-2)} + \cdots + u_n' y_n^{(n-2)}$$

The parameters commute out, with a little "dirt":

$$f = u_1 P(D) y_1 + u_2 P(D) y_2 + \cdots + u_n P(D) y_n + u_1' y_1^{(n-1)} + u_2' y_2^{(n-1)} + \cdots + u_n' y_n^{(n-1)}.$$

But $P(D)y_j = 0$, therefore

$$f = u_1' y_1^{(n-1)} + u_2' y_2^{(n-1)} + \cdots + u_n' y_n^{(n-1)}.$$

This, with the constraints, gives a linear system in the u'_j . This much can always be solved; in fact, combining Cramer's rule with the Wronskian,

$$u'_j = (-1)^{n+j} \frac{W(y_1,\ldots,y_{j-1},y_{j+1}\ldots,y_n)_{\binom{0}{f}}}{W(y_1,y_2,\ldots,y_n)}.$$

The rest is a matter of integrating u'_j.

The particular solution is not unique; $y_p + c_1y_1 + \cdots + c_ny_n$ also satisfies the ODE for any set of constants c_j.

Example: Suppose $y'' - 4y' + 5y = sin(kx)$. We take the solution basis found above $\{e^{(2+i)x}, e^{(2-i)x}\}$.

$$\begin{aligned} W &= \begin{vmatrix} e^{(2+i)x} & e^{(2-i)x} \\ (2+i)e^{(2+i)x} & (2-i)e^{(2-i)x} \end{vmatrix} \\ &= e^{4x} \begin{vmatrix} 1 & 1 \\ 2+i & 2-i \end{vmatrix} \\ &= -2ie^{4x} \end{aligned}$$

$$\begin{aligned} u'_1 &= \frac{1}{W} \begin{vmatrix} 0 & e^{(2-i)x} \\ \sin(kx) & (2-i)e^{(2-i)x} \end{vmatrix} \\ &= -\frac{i}{2}\sin(kx)e^{(-2-i)x} \end{aligned}$$

$$\begin{aligned} u'_2 &= \frac{1}{W} \begin{vmatrix} e^{(2+i)x} & 0 \\ (2+i)e^{(2+i)x} & \sin(kx) \end{vmatrix} \\ &= \frac{i}{2}\sin(kx)e^{(-2+i)x}. \end{aligned}$$

Using the list of integrals of exponential functions,

$$u_1 = -\frac{i}{2}\int \sin(kx)e^{(-2-i)x}\,dx$$

$$= \frac{ie^{(-2-i)x}}{2(3+4i+k^2)}\big((2+i)\sin(kx) + k\cos(kx)\big)$$

$$u_2 = \frac{i}{2}\int \sin(kx)e^{(-2+i)x}\,dx$$

$$= \frac{ie^{(i-2)x}}{2(3-4i+k^2)}\big((i-2)\sin(kx) - k\cos(kx)\big).$$

And so,

$$y_p = \frac{i}{2(3+4i+k^2)}((2+i)\sin(kx)+k\cos(kx)) + \frac{i}{2(3-4i+k^2)}((i-2)\sin(kx)-k\cos(kx))$$
$$= \frac{(5-k^2)\sin(kx)+4k\cos(kx)}{(3+k^2)^2+16}.$$

(u_1 and u_2 had factors that canceled y_1 and y_2; that is typical.)

For interest's sake, this ODE has a physical interpretation as a driven damped harmonic oscillator; y_p represents the steady state, and $c_1y_1 + c_2y_2$ is the transient.

First-order Linear ODEs

Example

$$y' + 3y = 2$$

with the initial condition

$$f(0) = 2.$$

Using the general solution method:

$$f = e^{-3x}\left(\int 2e^{3x}dx + \kappa\right).$$

The integration is done from 0 to x, giving:

$$f = e^{-3x}\left(2/3\left(e^{3x} - e^{0}\right) + \kappa\right).$$

Then we can reduce to:

$$f = 2/3\left(1 - e^{-3x}\right) + e^{-3x}\kappa.$$

For a first-order linear ODE, with coefficients that may or may not vary with x:

$$y'(x) + p(x)y(x) = r(x)$$

Then,

$$y = e^{-a(x)}\left(\int r(x)e^{a(x)}dx + \kappa\right)$$

Where κ is the constant of integration, and

$$a(x) = \int p(x)dx.$$

This proof comes from Jean Bernoulli.

Let,

$$y' + py = r$$

Suppose for some unknown functions $u(x)$ and $v(x)$ that $y=uv$.

Then,

$$y' = u'v + uv'$$

Substituting into the differential equation,

$$u'v + uv' + puv = r$$

Now, the most important step: Since the differential equation is linear we can split this into two independent equations and write,

$$u'v + puv = 0$$
$$uv' = r$$

Since v is not zero, the top equation becomes

$$u' + pu = 0$$

The solution of this is

$$u = e^{-\int pdx}$$

Substituting into the second equation

$$v = \int re^{\int pdx} + C$$

Since $y=uv$, for arbitrary constant C

$$y = e^{-\int pdx}\left(\int re^{\int pdx} + C\right)$$

As an illustrative example, consider a first order differential equation with constant coefficients:

$$\frac{dy}{dx} + by = 1.$$

This equation is particularly relevant to first order systems such as RC circuits and mass-damper systems.

In this case, $p(x) = b$, $r(x) = 1$.

Hence its solution is

$$y(x) = e^{-bx}\left(e^{bx}/b + C\right) = 1/b + Ce^{-bx}.$$

Method of Undetermined Coefficients

The method of undetermined coefficients (MoUC), is useful in finding solution for y_p. Given the ODE $P(D)y = f(x)$, find another differential operator $A(D)$ such that $A(D)$ $f(x) = 0$. This operator is called the annihilator, and thus the method of undetermined coefficients is also known as the annihilator method. Applying $A(D)$ to both sides of the ODE gives a homogeneous ODE $\left(A(D)P(D)\right)y = 0$ for which we find a solution basis $\{y_1,\ldots,y_n\}$ as before. Then the original nonhomogeneous ODE is used to construct a system of equations restricting the coefficients of the linear combinations to satisfy the ODE.

Undetermined coefficients is not as general as variation of parameters in the sense that an annihilator does not always exist.

Example: Given $y'' - 4y' + 5y = sin(kx), P(D) = D^2 - 4D + 5$. The simplest annihilator of $\sin(kx)$ is $A(D) = D^2 + k^2$. The zeros of $A(z)P(z)$ are $\{2+i, 2-i, ik, -ik\}$, so the solution basis of $A(D)P(D)$ is $\{y_1, y_2, y_3, y_4\} = \{e^{(2+i)x}, e^{(2-i)x}, e^{ikx}, e^{-ikx}\}$.

Setting $y = c_1y_1 + c_2y_2 + c_3y_3 + c_4y_4$ we find

$$\begin{aligned}
\sin(kx) &= P(D)y \\
&= P(D)(c_1y_1 + c_2y + c_3y_3 + c_4y_4) \\
&= c_1P(D)y_1 + c_2P(D)y_2 + c_3P(D)y_3 + c_4P(D)y_4 \\
&= 0 + 0 + c_3\left(-k^2 - 4ik + 5\right)y_3 + c_4\left(-k^2 + 4ik + 5\right)y_4 \\
&= c_3\left(-k^2 - 4ik + 5\right)\left(\cos(kx) + i\sin(kx)\right) + c_4\left(-k^2 + 4ik + 5\right)\left(\cos(kx) - i\sin(kx)\right)
\end{aligned}$$

giving the system

$$\begin{aligned}
i &= \left(k^2 + 4ik - 5\right)c_3 + \left(-k^2 + 4ik + 5\right)c_4 \\
0 &= \left(k^2 + 4ik - 5\right)c_3 + \left(k^2 - 4ik - 5\right)c_4
\end{aligned}$$

which has solutions

$$c_3 = \frac{i}{2(k^2 + 4ik - 5)}, \; c_4 = \frac{i}{2(-k^2 + 4ik + 5)}$$

Giving the solution set

$$\begin{aligned} y &= c_1y_1 + c_2y_2 + \frac{i}{2(k^2+4ik-5)}y_3 + \frac{i}{2(-k^2+4ik+5)}y_4 \\ &= c_1y_1 + c_2y_2 + \frac{4k\cos(kx)-(k^2-5)\sin(kx)}{(k^2+4ik-5)(k^2-4ik-5)} \\ &= c_1y_1 + c_2y_2 + \frac{4k\cos(kx)+(5-k^2)\sin(kx)}{k^4+6k^2+25}. \end{aligned}$$

Method of Variation of Parameters

The general solution to a non-homogeneous, linear differential equation $y''(x) + p(x)y'(x) + q(x)y(x) = g(x)$ can be expressed as the sum of the general solution $y_h(x)$ to the corresponding homogenous, linear differential equation $y''(x) + p(x)y'(x) + q(x)y(x) = 0$ and any one solution $y_p(x)$ to $y_p(x) to\, y''(x) + p(x)y'(x) + q(x)y(x) = g(x)$.

Like the method of undetermined coefficients, described above, the method of variation of parameters is a method for finding one solution to $y''(x) + p(x)y'(x) + q(x)y(x) = g(x)$, having already found the general solution to $y''(x) + p(x)y'(x) + q(x)y(x) = 0$. Unlike the method of undetermined coefficients, which fails except with certain specific forms of $g(x)$, the method of variation of parameters will always work; however, it is significantly more difficult to use.

For a second-order equation, the method of variation of parameters makes use of the following fact:

Fact

Let $p(x), q(x)$, and $g(x)$ be functions, and let $y_1(x)$ and $y_2(x)$ be solutions to the homogeneous, linear differential equation $y''(x) + p(x)y'(x) + q(x)y(x) = 0$. Further, let $u(x)$ and $v(x)$ be functions such that $u'(x)y_1(x) + v'(x)y_2(x) = 0$ and $u'(x)y_1'(x) + v'(x)y_2'(x) = g(x)$ for all x, and define $y_p(x) = u(x)y_1(x) + v(x)y_2(x)$. Then $y_p(x)$ is a solution to the non-homogeneous, linear differential equation $y''(x) + p(x)y'(x) + q(x)y(x) = g(x)$.

Proof

$$y_p(x) = u(x)y_1(x) + v(x)y_2(x)$$

$$\begin{aligned} y_{p'}(x) &= u'(x)y_1(x) + u(x)y_{1'}(x) + v'(x)y_2(x) + v(x)y_{2'}(x) \\ &= 0 + u(x)y_{1'}(x) + v(x)y_{2'}(x) \end{aligned}$$

$$\begin{aligned} y_{p''}(x) &= u'(x)y_{1'}(x) + u(x)y_{1''}(x) + v'(x)y_{2'}(x) + v(x)y_{2''}(x) \\ &= g(x) + u(x)y_{1''}(x) + v(x)y_{2''}(x) \end{aligned}$$

$$y_{p''}(x) + p(x)y_p'(x) + q(x)y_p(x) = g(x) + u(x)y_{1''}(x) + v(x)y_{2''}(x) + p(x)u(x)y_{1'}(x)$$

$$+ p(x)v(x)y_{2'}(x) + q(x)u(x)y_1(x) + q(x)v(x)y_2(x)$$

Usage

To solve the second-order, non-homogeneous, linear differential equation $y''(x) + p(x)y'(x) + q(x)y(x) = g(x)$ using the method of variation of parameters, use the following steps:

1. Find the general solution to the corresponding homogeneous equation $y''(x) + p(x)y'(x) + q(x)y(x) = 0$. Specifically, find two linearly independent solutions $y_1(x)$ and $y_2(x)$.
2. Since $y_1(x)$ and $y_2(x)$ are linearly independent solutions, their Wronskian $y_1(x)y_2'(x) - y_1'(x)y_2(x)$ is nonzero, so we can compute $-(g(x)y_2(x)) / (y_1(x)y_2'(x) - y_1'(x)y_2(x))$ and $(g(x)y_1(x)) / (y_1(x)y_2'(x) - y_1'(x)y_2(x))$. If the former is equal to $u'(x)$ and the latter to $v'(x)$, then u and v satisfy the two constraints given above: that $u'(x)y_1(x) + v'(x)y_2(x) = 0$ and that $u'(x)y_1'(x) + v'(x)y_2'(x) = g(x)$. We can tell this after multiplying by the denominator and comparing coefficients.
3. Integrate $-(g(x)y_2(x)) / (y_1(x)y_2'(x) - y_1'(x)y_2(x))$ and $(g(x)y_1(x)) / (y_1(x)y_2'(x) - y_1'(x)y_2(x))$ to obtain $u(x)$ and $v(x)$, respectively. (Note that we only need one choice of u and v, so there is no need for constants of integration.)
4. Compute $y_p(x) = u(x)y_1(x) + v(x)y_2(x)$. The function y_p is one solution of $y''(x) + p(x)y'(x) + q(x)y(x) = g(x)$.
5. The general solution is $c_1y_1(x) + c_2y_2(x) + y_p(x)$, where c_1 and c_2 are arbitrary constants.

Higher-order Equations

The method of variation of parameters can also be used with higher-order equations. For example, if $y_1(x), y_2(x)$, and $y_3(x)$ are linearly independent solutions $''(x) + p(x)y''(x) + q(x)y'(x) + r(x)y(x) = 0$, then there exist functions $u(x), v(x)$, and $w(x)$ such that,

$u'(x)y_1(x) + v'(x)y_2(x) + w'(x)y_3(x) = 0, u'(x)y_1'(x) + v'(x)y_2'(x) + w'(x)y_3'(x) = 0,$ and $u'(x)y_1''(x) + v'(x)y_2''(x) + w'(x)y_3''(x) = g(x).$ Having found such functions (by solving algebraically for $u'(x), v'(x),$ *and* $w'(x)$, then integrating each), we have $y_p(x) = u(x)y_1(x) + v(x)y_2(x) + w(x)y_3(x)$, one solution to the equation $y'''(x) + p(x)y''(x) + q(x)y'(x) + r(x)y(x) = g(x).$

Example

Solve the previous example, $y''+y= \sec x$ Recall $\sec x = \frac{1}{\cos x} = f$. LHS has root of $r = \pm i$ that yield $y_c = C_1 cosx + C_2 sinx$, (so $y_1 = cosx, y_2 = sinx$) and its derivatives

$$\begin{cases} \dot{u} = \frac{-y_2 f}{W} = \frac{-\sin x}{\cos x} = \tan x \\ \dot{v} = \frac{y_1 f}{W} = \frac{\cos x}{\cos x} = 1 \end{cases}$$

where the Wronskian

$$W(y_1, y_2 : x) = \begin{vmatrix} \cos x & \sin x \\ -\sin x & \cos x \end{vmatrix} = 1$$

were computed in order to seek solution to its derivatives.

Upon integration,

$$\begin{cases} u = -\int \tan x dx = -\ln|\sec x| + C \\ v = \int 1 dx = x + C \end{cases}$$

Computing y_p and y_G,

$$y_p = f = uy_1 + vy_2 = \cos x \ln|\cos x| + x\sin x$$
$$y_G = y_c + y_p = C_1 \cos x + C_2 \sin x + x \sin x + \cos x \ln(\cos x)$$

Linear ODEs with Variable Coefficients

A linear ODE of order n with variable coefficients has the general form

$$p_n(x)y^{(n)}(x) + p_{n-1}(x)y^{(n-1)}(x) + \ldots + p_0(x)y(x) = r(x).$$

Examples

A particular simple example is the Cauchy-Euler equation often used in engineering

$$x^n y^{(n)}(x) + a_{n-1}x^{n-1}y^{(n-1)}(x) + \ldots + a_0 y(x) = 0.$$

Partial Differential Equation

A partial differential equation (or briefly a PDE) is a mathematical equation that involves two or more independent variables, an unknown function (dependent on those variables), and partial derivatives of the unknown function with respect to the independent variables. The order of a partial differential equation is the order of the highest derivative involved. A solution (or a particular solution) to a partial differential equation is a function that solves the equation or, in other words, turns it into an identity when substituted into the equation. A solution is called general if it contains all particular solutions of the equation concerned.

The term exact solution is often used for second- and higher-order nonlinear PDEs to denote a particular solution.

Partial differential equations are used to mathematically formulate, and thus aid the solution of, physical and other problems involving functions of several variables, such as the propagation of heat or sound, fluid flow, elasticity, electrostatics, electrodynamics, etc.

First-Order Partial Differential Equations

General Form of First-Order Partial Differential Equation

A first-order partial differential equation with nindependent variables has the general form

$$F\left(x_1, x_2, \ldots, x_n, w, \frac{\partial w}{\partial x_1}, \frac{\partial w}{\partial x_2}, \ldots, \frac{\partial w}{\partial xn}\right) = 0,$$

where $w = w(x_1, x_2, \ldots, x_n)$ is the unknown function and $F(\ldots)$ is a given function.

Quasilinear Equations, Characteristic System and General Solution

General form of first-order quasilinear PDE

A first-order quasilinear partial differential equation with two independent variables has the general form,

$$f(x, y, w)\frac{\partial w}{\partial x} + g(x, y, w)\frac{\partial w}{\partial y} = h(x, y, w).$$

Such equations are encountered in various applications (continuum mechanics, gas dynamics, hydrodynamics, heat and mass transfer, wave theory, acoustics, multiphase flows, chemical engineering, etc.).

If the functions f, g, and h are independent of the unknown w, then equation $f(x,y,w)\frac{\partial w}{\partial x}+g(x,y,w)\frac{\partial w}{\partial y}=h(x,y,w).$ is called linear characteristic system.

General solution:

The system of ordinary differential equations,

$$\frac{dx}{f(x,y,w)}=\frac{dy}{g(x,y,w)}=\frac{dw}{h(x,y,w)}$$

is known as the characteristic system of equation $f(x,y,w)\frac{\partial w}{\partial x}+g(x,y,w)\frac{\partial w}{\partial y}=h(x,y,w).$

Suppose that two independent particular solutions of this system have been found in the form,

$$u_1(x,y,w)=C_1, \qquad u_2(x,y,w)=C_2,$$

where C_1 and C_2 are arbitrary constants; such particular solutions are known as integrals of system. Then the general solution to equation $f(x,y,w)\frac{\partial w}{\partial x}+g(x,y,w)\frac{\partial w}{\partial y}=h(x,y,w).$ can be written as

$$\Phi(u_1,u_2)=0,$$

where Φ is an arbitrary function of two variables. With equation $\Phi(u_1,u_2)=0,$ solved for u_2, one often specifies the general solution in the form $u_2=\Psi(u_1)$, where $\Psi(u)$ is an arbitrary function of one variable.

If $h(x,y,w)\equiv 0,$ then $w=C_2$ can be used as the second integral in above equation.

Example. Consider the linear equation,

$$\frac{\partial w}{\partial x}+a\frac{\partial w}{\partial y}=b.$$

The associated characteristic system of ordinary differential equations

$$\frac{dx}{1}=\frac{dy}{a}=\frac{dw}{b}$$

has two integrals

$$y-ax=C_1, \qquad w-bx=C_2.$$

Therefore, the general solution to this PDE can be written as $w - bx = \Psi(y - ax)$, or

$$w = bx + \Psi(y - ax),$$

where $\Psi(z)$ is an arbitrary function.

Cauchy Problem: Two Formulations and Solving the Cauchy Problem

Generalized Cauchy Problem

Generalized Cauchy problem: find a solution $w = w(x, y)$ to equation $f(x,y,w)\frac{\partial w}{\partial x} + g(x,y,w)\frac{\partial w}{\partial y} = h(x,y,w)$. satisfying the initial conditions

$$x = \varphi_1(\xi), \quad y = \varphi_2(\xi), \quad w = \varphi_3(\xi),$$

where ξ is a parameter $(\alpha \le \xi \le \beta)$ and the $\varphi_k(\xi)$ are given functions.

Geometric interpretation: find an integral surface of equation $f(x,y,w)\frac{\partial w}{\partial x} + g(x,y,w)\frac{\partial w}{\partial y} = h(x,y,w)$ passing through the line defined parametrically by equation $x = \varphi_1(\xi), \quad y = \varphi_2(\xi), \quad w = \varphi_3(\xi)$.

Classical Cauchy Problem

Classical Cauchy problem: find a solution $w = w(x, y)$ of equation $f(x,y,w)\frac{\partial w}{\partial x} + g(x,y,w)\frac{\partial w}{\partial y} = h(x,y,w)$. satisfying the initial condition

$$w = \varphi(y) \quad at \quad x = 0,$$

where $\varphi(y)$ is a given function.
It is often convenient to represent the classical Cauchy problem as a generalized Cauchy problem by rewriting condition $w = \varphi(y) \quad at \quad x = 0$, in the parametric form $x = 0, \ y = \xi, \quad w = \varphi(\xi)$.

Existence and Uniqueness Theorem

If the coefficients f, g, and h of equation $f(x,y,w)\frac{\partial w}{\partial x} + g(x,y,w)\frac{\partial w}{\partial y} = h(x,y,w)$. and the functions φ_k in equation above are continuously differentiable with respect to each of their arguments and if the inequalities $f\varphi'_2 - g\varphi'_1 \ne 0$ and $(\varphi'_1)^2 + (\varphi'_2)^2 \ne 0$ hold along the curve, then there is a unique solution to the Cauchy problem (in a neighborhood of the curve).

Procedure of Solving the Cauchy Problem

The procedure for solving the Cauchy problem involves several steps. First, two independent integrals of the characteristic system are determined. Then, to find the constants of integration C_1 and C_2, the initial data must be substituted into the integrals $u_1(x,y,w)=C_1, \quad u_2(x,y,w)=C_2$, to obtain

$$u_1(\varphi_1(\xi),\varphi_2(\xi),\varphi_3(\xi))=C_1, \quad u_2(\varphi_1(\xi),\varphi_2(\xi),\varphi_3(\xi))=C_2.$$

Eliminating C_1 and C_2 from $u_1(x,y,w)=C_1, \quad u_2(x,y,w)=C_2$, and above equation yields

$$u_1(x,y,w)=u_1(\varphi_1(\xi),\varphi_2(\xi),\varphi_3(\xi)),$$
$$u_2(x,y,w)=u_2(\varphi_1(\xi),\varphi_2(\xi),\varphi_3(\xi)).$$

Formulas are a parametric form of the solution to the Cauchy problem. In some cases, one may succeed in eliminating the parameter ξ from relations, thus obtaining the solution in an explicit form.

Second-Order Partial Differential Equations

Linear, Semilinear and Nonlinear Second-Order PDEs

Linear second-order PDEs and their properties. Principle of linear superposition.

A second-order linear partial differential equation with two independent variables has the form

$$a(x,y)\frac{\partial^2 w}{\partial x^2}+2b(x,y)\frac{\partial^2 w}{\partial x \partial y}+c(x,y)\frac{\partial^2 w}{\partial y^2}=\alpha(x,y)\frac{\partial w}{\partial x}+\beta(x,y)\frac{\partial w}{\partial y}+\gamma(x,y)w+\delta(x,y).$$

If $\delta \equiv 0$, equation above is a homogeneous linear equation, and if $\delta \not\equiv 0$, it is a nonhomogeneous linear equation. The functions $a(x,y),b(x,y), \ldots,\gamma(x,y),\delta(x,y)$ are called coefficients of equation above.

Some properties of a homogeneous linear equation (with $\delta \equiv 0$):

1. A homogeneous linear equation has a particular solution $w=0$.
2. The principle of linear superposition holds; namely, if $w_1(x,y), w_2(x,y), \ldots, w_n(x,y)$ are particular solutions to homogeneous linear equation, then the function $A_1 w_1(x,y)+A_2 w_2(x,y)+\cdots+A_n w_n(x,y)$, where $A_1, A_2, \ldots, A_n$ are arbitrary numbers is also an exact solution to that equation.
3. Suppose equation
$$a(x,y)\frac{\partial^2 w}{\partial x^2}+2b(x,y)\frac{\partial^2 w}{\partial x \partial y}+c(x,y)\frac{\partial^2 w}{\partial y^2}=\alpha(x,y)\frac{\partial w}{\partial x}+\beta(x,y)\frac{\partial w}{\partial y}+\gamma(x,y)w+\delta(x,y).$$
has a particular solution $\widetilde{w}=\widetilde{w}(x,y;\mu)$ that depends on a parameter μ, and the coefficients of the linear differential equation are independent of μ (but can

depend on x and y). Then, by differentiating $\widetilde{w}$ with respect to μ, one obtains other solutions to the equation, $\frac{\partial \widetilde{w}}{\partial \mu}, \frac{\partial^2 \widetilde{w}}{\partial \mu^2}, \ldots, \frac{\partial^k \widetilde{w}}{\partial \mu^k}, \ldots$

4. Let $\tilde{w} = \tilde{w}(x, y; \mu)$ be a particular solution as described in property 3. Multiplying $\tilde{w}$ by an arbitrary function $\varphi(\mu)$ and integrating the resulting expression with respect to μ over some interval $[\mu_1, \mu_2]$, one obtains a new function $\int_{\mu_1}^{\mu_2} \widetilde{w}(x, y; \mu)\varphi(\mu)d\mu$, which is also a solution to the original homogeneous linear equation.

5. Suppose the coefficients of the homogeneous linear equation $a(x,y)\frac{\partial^2 w}{\partial x^2} + 2b(x,y)\frac{\partial^2 w}{\partial x \partial y} + c(x,y)\frac{\partial^2 w}{\partial y^2} = \alpha(x,y)\frac{\partial w}{\partial x} + \beta(x,y)\frac{\partial w}{\partial y} + \gamma(x,y)w + \delta(x,y).$ are independent of x. Then: (i) there is a particular solution of the form $w = e^{\lambda x}u(y),$, where λ is an arbitrary number and $u(y)$ is determined by a linear second-order ordinary differential equation, and (ii) differentiating any particular solution with respect to x also results in a particular solution to equation $a(x,y)\frac{\partial^2 w}{\partial x^2} + 2b(x,y)\frac{\partial^2 w}{\partial x \partial y} + c(x,y)\frac{\partial^2 w}{\partial y^2} = \alpha(x,y)\frac{\partial w}{\partial x} + \beta(x,y)\frac{\partial w}{\partial y} + \gamma(x,y)w + \delta(x,y).$

Properties 2–5 are widely used for constructing solutions to problems governed by linear PDEs.

Examples of particular solutions to linear PDEs can be found in the subsections Heat equation and Laplace equation below.

Semilinear and Nonlinear Second-order PDEs

A second-order semilinear partial differential equation with two independent variables has the form,

$$a(x,y)\frac{\partial^2 w}{\partial x^2} + 2b(x,y)\frac{\partial^2 w}{\partial x \partial y} + c(x,y)\frac{\partial^2 w}{\partial y^2} = F\left(x, y, w, \frac{\partial w}{\partial x}, \frac{\partial w}{\partial x}\right)$$

In the general case, a second-order nonlinear partial differential equation with two independent variables has the form,

$$F\left(x, y, w, \frac{\partial w}{\partial x}, \frac{\partial w}{\partial y}, \frac{\partial^2 w}{\partial x^2}, \frac{\partial^2 w}{\partial x \partial y}, \frac{\partial^2 w}{\partial y^2}\right) = 0.$$

The classification and the procedure for reducing linear and semilinear equations of the form above equations to a canonical form are only determined by the left-hand side of the equations.

Some Linear Equations Encountered in Applications

Three basic types of linear partial differential equations are distinguished—parabolic, hyperbolic, and elliptic. The solutions of the equations pertaining to each of the types have their own characteristic qualitative differences.

Heat Equation (A Parabolic Equation)

1. The simplest example of a parabolic equation is the heat equation

$$\frac{\partial w}{\partial t} - \frac{\partial^2 w}{\partial x^2} = 0,$$

where the variables t and x play the role of time and a spatial coordinate, respectively. Note that equation $\frac{\partial w}{\partial t} - \frac{\partial^2 w}{\partial x^2} = 0$, contains only one highest derivative term.

Equation $\frac{\partial w}{\partial t} - \frac{\partial^2 w}{\partial x^2} = 0$, is often encountered in the theory of heat and mass transfer. It describes one-dimensional unsteady thermal processes in quiescent media or solids with constant thermal diffusivity. A similar equation is used in studying corresponding one-dimensional unsteady mass-exchange processes with constant diffusivity.

2. The heat equation $\frac{\partial w}{\partial t} - \frac{\partial^2 w}{\partial x^2} = 0$, has infinitely many particular solutions (this fact is common to many PDEs); in particular, it admits solutions

$$w(x,t) = A(x^2 + 2t) + B,$$
$$w(x,t) = Aexp(\mu^2 t \pm \mu x) + B,$$
$$w(x,t) = A\frac{1}{\sqrt{t}}exp\left(-\frac{x^2}{4t}\right) + B,$$
$$w(x,t) = Aexp(-\mu^2 t)cos(\mu x + B) + C,$$
$$w(x,t) = Aexp(-\mu x)cos(\mu x - 2\mu^2 t + B) + C,$$

where A, B, C, and μ are arbitrary constants.

Wave Equation (A Hyperbolic Equation)

1. The simplest example of a hyperbolic equation is the wave equation

$$\frac{\partial^2 w}{\partial t^2} - \frac{\partial^2 w}{\partial x^2} = 0,$$

Where the variables t and x play the role of time and the spatial coordinate, respectively. The highest derivative terms in equation $\frac{\partial^2 w}{\partial t^2} - \frac{\partial^2 w}{\partial x^2} = 0$, differ in sign.

This equation is also known as the equation of vibration of a string. It is often encountered in elasticity, aerodynamics, acoustics, and electrodynamics.

2. The general solution of the wave equation $\frac{\partial^2 w}{\partial t^2} - \frac{\partial^2 w}{\partial x^2} = 0,$

$$w = \varphi(x+t) + \psi(x-t).$$

Where $\varphi(x)$ and $\psi(x)$ are arbitrary twice continuously differentiable functions. This solution has the physical interpretation of two traveling waves of arbitrary shape that propagate to the right and to the left along the x -axis with a constant speed equal to 1.

Laplace Equation (An Elliptic Equation)

1. The simplest example of an elliptic equation is the Laplace equation

$$\frac{\partial^2 w}{\partial x^2} + \frac{\partial^2 w}{\partial y^2} = 0,$$

where x and y play the role of the spatial coordinates. Note that the highest derivative terms in equation $\frac{\partial^2 w}{\partial x^2} + \frac{\partial^2 w}{\partial y^2} = 0$, have like signs. The Laplace equation is often written briefly as $\Delta w = 0$, where Δ is the Laplace operator.

The Laplace equation is often encountered in heat and mass transfer theory, fluid mechanics, elasticity, electrostatics, and other areas of mechanics and physics. For example, in heat and mass transfer theory, this equation describes steady-state temperature distribution in the absence of heat sources and sinks in the domain under study.

A solution to the Laplace equation $\frac{\partial^2 w}{\partial x^2} + \frac{\partial^2 w}{\partial y^2} = 0$, is called a harmonic function.

2. The particular solutions of the Laplace equation $\frac{\partial^2 w}{\partial x^2} + \frac{\partial^2 w}{\partial y^2} = 0$:

$$w(x,y) = Ax + By + C,$$

$$w(x,y) = A(x^2 - y^2) + Bxy,$$

$$w(x,y) = \frac{Ax + By}{x^2 + y^2} + C,$$

$$w(x,y) = (A \sin h\, \mu x + B \cosh \mu x)(C \cos \mu y + D \sin \mu y),$$

$$w(x,y) = (A \cos \mu x + B \sin \mu x)(C \sinh \mu y + D \cosh \mu y),$$

Where $A, B, C, D,$ and μ are arbitrary constants.

A fairly general method for constructing solutions to the Laplace equation $\frac{\partial^2 w}{\partial x^2} + \frac{\partial^2 w}{\partial y^2} = 0$, involves the following. Let $f(z) = u(x,y) + iv(x,y)$ be any analytic function of the complex variable $z = x + iy$ (u and v are real functions of the real variables x and y; $i^2 = -1$. Then the real and imaginary parts of f both satisfy the Laplace equation,

$$\Delta u = 0, \qquad \Delta v = 0.$$

Thus, by specifying analytic functions $f(z)$ and taking their real and imaginary parts, one obtains various solutions of the Laplace equation $\frac{\partial^2 w}{\partial x^2} + \frac{\partial^2 w}{\partial y^2} = 0,$

Classification of Second-Order Partial Differential Equations

Types of Equations

Any semilinear partial differential equation of the second-order with two independent variables can be reduced, by appropriate manipulations, to a simpler equation that has one of the three highest derivative combinations specified above in examples.

Given a point (x, y) equation $a(x,y)\frac{\partial^2 w}{\partial x^2} + 2b(x,y)\frac{\partial^2 w}{\partial x \partial y} + c(x,y)\frac{\partial^2 w}{\partial y^2} = F\left(x, y, w, \frac{\partial w}{\partial x}, \frac{\partial w}{\partial x}\right)$ is said to be,

$$\begin{aligned} &\textit{parabolic} \quad \textit{if } b^2 - ac = 0, \\ &\textit{hyperbolic} \quad \textit{if } b^2 - ac > 0 \\ &\textit{elliptic} \quad \textit{if } b^2 - ac < 0 \end{aligned}$$

at this point.

Characteristic Equations

In order to reduce equation $a(x,y)\frac{\partial^2 w}{\partial x^2} + 2b(x,y)\frac{\partial^2 w}{\partial x \partial y} + c(x,y)\frac{\partial^2 w}{\partial y^2} = F\left(x, y, w, \frac{\partial w}{\partial x}, \frac{\partial w}{\partial x}\right)$ to a canonical form, one should first write out the characteristic equation

$$a(dy)^2 - 2b\, dx\, dy + c(dx)^2 = 0,$$

which with $a \neq 0$ splits into two equations

$$a\, dy - (b + \sqrt{b^2 - ac})\, dx = 0.$$

and

$$a\, dy - (b - \sqrt{b^2 - ac})\, dx = 0.$$

and then find their general integrals.

If $a \equiv 0$,, the simpler equations

$$\begin{aligned} dx &= 0, \\ 2b\, dy - c\, dx &= 0 \end{aligned}$$

should be used instead of $a\,dy-(b+\sqrt{b^2-ac})\,dx=0$ and $a\,dy-(b-\sqrt{b^2-ac})\,dx=0$. The first equation has the obvious general solution $x=C$.

Canonical form of Parabolic Equations

$(case\, b^2-ac=0)$

In this case, equations $a\,dy-(b+\sqrt{b^2-ac})\,dx=0$ and $a\,dy-(b-\sqrt{b^2-ac})\,dx=0$ coincide and have a common general integral,

$$u(x,y)=C.$$

By passing from x,y to new independent variables ξ,η in accordance with the relations

$$\xi=u(x,y), \qquad \eta=\eta(x,y),$$

where $\eta=\eta(x,y)$ is any twice differentiable function that satisfies the condition of nondegeneracy of the Jacobian $\frac{D(\xi,\eta)}{D(x,y)}$ in the given domain, one reduces equation $a(x,y)\frac{\partial^2 w}{\partial x^2}+2b(x,y)\frac{\partial^2 w}{\partial x\partial y}+c(x,y)\frac{\partial^2 w}{\partial y^2}=F\left(x,y,w,\frac{\partial w}{\partial x},\frac{\partial w}{\partial x}\right)$ to the canonical form

$$\frac{\partial^2 w}{\partial \eta^2}=F_1\left(\xi,\eta,w,\frac{\partial w}{\partial \xi},\frac{\partial w}{\partial \eta}\right).$$

As η, one can take $\eta=x\, or\, \eta=y$.

It is apparent that the transformed equation $\frac{\partial^2 w}{\partial \eta^2}=F_1\left(\xi,\eta,w,\frac{\partial w}{\partial \xi},\frac{\partial w}{\partial \eta}\right)$. has only one highest-derivative term, just as the heat equation $\frac{\partial w}{\partial t}-\frac{\partial^2 w}{\partial x^2}=0,$

Two Canonical forms of Hyperbolic Equations

$(case\, b^2-ac>0)$

1. The general integrals

$$u_1(x,y)=C_1, \qquad u_2(x,y)=C_2$$

of equations $a\,dy-(b+\sqrt{b^2-ac})\,dx=0$ and $a\,dy-(b-\sqrt{b^2-ac})\,dx=0$ are real and different. These integrals determine two different families of real characteristics.

By passing from x,y to new independent variables ξ,η in accordance with the relations

$$\xi=u_1(x,y), \qquad \eta=u_2(x,y),$$

one reduces equation $a(x,y)\frac{\partial^2 w}{\partial x^2}+2b(x,y)\frac{\partial^2 w}{\partial x\partial y}+c(x,y)\frac{\partial^2 w}{\partial y^2}=F\left(x,y,w,\frac{\partial w}{\partial x},\frac{\partial w}{\partial x}\right)$ to $\frac{\partial^2 w}{\partial\xi\,\partial\eta}=F_2\left(\xi,\eta,w,\frac{\partial w}{\partial\xi},\frac{\partial w}{\partial\eta}\right)$.

This is the so-called first canonical form of a hyperbolic equation.

2. The transformation

$$\xi=t+z,\qquad \eta=t-z$$

brings the above equation to another canonical form,

$$\frac{\partial^2 w}{\partial t^2}-\frac{\partial^2 w}{\partial z^2}=F_3\left(t,z,w,\frac{\partial w}{\partial t},\frac{\partial w}{\partial z}\right),$$

where $F_3=4F_2$. This is the so-called second canonical form of a hyperbolic equation. Apart from notation, the left-hand side of the last equation coincides with that of the wave equation $\frac{\partial^2 w}{\partial t^2}-\frac{\partial^2 w}{\partial x^2}=0$.

Canonical form of Elliptic Equations

(case $b^2-ac<0$)

In this case the general integrals of equations $a\,dy-(b+\sqrt{b^2-ac})\,dx=0$ and $a\,dy-(b-\sqrt{b^2-ac})\,dx=0$ are complex conjugates; these determine two families of complex characteristics.

Let the general integral of equation $a\,dy-(b+\sqrt{b^2-ac})\,dx=0$ have the form

$$u_1(x,y)+iu_2(x,y)=C,\qquad i^2=-1,$$

where $u_1(x,y)$ and $u_2(x,y)$ are real-valued functions.

By passing from x,y to new independent variables ξ,η in accordance with the relations

$$\xi=u_1(x,y),\qquad \eta=u_2(x,y),$$

one reduces equation above to the canonical form

$$\frac{\partial^2 w}{\partial\xi^2}+\frac{\partial^2 w}{\partial\eta^2}=F_4(\xi,\eta,w,\frac{\partial w}{\partial\xi},\frac{\partial w}{\partial\eta}).$$

Apart from notation, the left-hand side of the last equation coincides with that of the Laplace equation $\frac{\partial^2 w}{\partial x^2}+\frac{\partial^2 w}{\partial y^2}=0$.

Basic Problems for PDEs of Mathematical Physics

Most PDEs of mathematical physics govern infinitely many qualitatively similar phenomena or processes. This follows from the fact that differential equations have, as a rule, infinitely many particular solutions. The specific solution that describes the physical phenomenon under study is separated from the set of particular solutions of the given differential equation by means of the initial and boundary conditions.

For simplicity and clarity, the basic problems of mathematical physics will be presented for the simplest linear equations $\frac{\partial w}{\partial t}-\frac{\partial^2 w}{\partial x^2}=0$, $\frac{\partial^2 w}{\partial t^2}-\frac{\partial^2 w}{\partial x^2}=0$, and $\frac{\partial^2 w}{\partial x^2}+\frac{\partial^2 w}{\partial y^2}=0$, only.

Cauchy Problem and Boundary Value Problems for Parabolic Equations

Cauchy problem $(t \geq 0, -\infty < x < \infty)$. Find a function w that satisfies heat equation $\frac{\partial w}{\partial t}-\frac{\partial^2 w}{\partial x^2}=0$, for $t>0$ and the initial condition $w=\varphi(x)\ \ at\ \ t=0$.

The solution of the Cauchy problem is

$$w(x,t)=\int_{-\infty}^{\infty}\varphi(\xi)E(x,\xi,t)d\xi$$

where $E(x,\xi,t)$ is the fundamental solution of the Cauchy problem,

$$E(x,\xi,t)=\frac{1}{2\sqrt{\pi a t}}exp\left[-\frac{(x-\xi)^2}{4at}\right].$$

In all boundary value problems (or initial-boundary value problems) below, it will be required to find a function w, in a domain $t \geq 0$,$x_1 \leq x \leq x_2$ $(-\infty < x_1 < x_2 < \infty)$, that satisfies the heat equation $\frac{\partial w}{\partial t}-\frac{\partial^2 w}{\partial x^2}=0$, for $t>0$ and the initial condition. In addition, all problems will be supplemented with some boundary conditions as given below.

First boundary value problem. The function $w(x,t)$ takes prescribed values on the boundary:

$$w=\psi_1(t) \quad at \quad x=x_1,$$
$$w=\psi_2(t) \quad at \quad x=x_2.$$

In particular, the solution to the first boundary value problem with $\psi_1(t)=\psi_2(t)\equiv 0, x_1=0$, and $x_2=l$ is expressed as

$$w(x,t)=\int_0^l \varphi(\xi)G(x,\xi,t)d\xi,$$

where the Green's function $G(x,\xi,t)$ is defined by the formulas

$$G(x,\xi,t)=\frac{2}{l}\sum_{n=1}^{\infty} sin\left(\frac{n\pi x}{l}\right) sin\left(\frac{n\pi \xi}{l}\right) exp\left(\frac{-an^2\pi^2 t}{l^2}\right)$$

$$=\frac{1}{2\sqrt{\pi at}}\sum_{n=-\infty}^{\infty}\left\{exp\left[-\frac{(x-\xi+2nl)^2}{4at}\right]-exp\left[-\frac{(x+\xi+2nl)^2}{4at}\right]\right\}$$

The first series converges rapidly at large t and the second series at small t.

Second boundary value problem. The derivatives of the function $w(x\ t)$ are prescribed on the boundary:

$$\frac{\partial w}{\partial x}=\psi_1(t)\ \ at\ \ x=x_1,$$

$$\frac{\partial w}{\partial x}=\psi_2(t)\ \ at\ \ x=x_2.$$

Third boundary value problem. A linear relationship between the unknown function and its derivatives are prescribed on the boundary:

$$\frac{\partial w}{\partial x}-k_1 w=\psi_1(t)\ \ at\ \ x=x_1,$$

$$\frac{\partial w}{\partial x}+k_2 w=\psi_2(t)\ \ at\ \ x=x_2.$$

Mixed boundary value problems. Conditions of different type, listed above, are set on the boundary of the domain in question, for example,

$$x \neq \psi_1(t)\qquad at\ \ x=x_1,$$

$$\frac{\partial w}{\partial x}=\psi_2(t)\ \ at\ \ x=x_2.$$

The boundary conditions are called homogeneous if $\psi_1(t)=\psi_2(t)\equiv 0$.

Solutions to the above initial-boundary value problems for the heat equation can be obtained by separation of variables (Fourier method) in the form of infinite series or by the method of integral transforms using the Laplace transform.

Cauchy problem and boundary value problems for hyperbolic equations

Cauchy problem $(t\geq 0, -\infty<x<\infty)$. Find a function w that satisfies the wave equation

$\frac{\partial^2 w}{\partial t^2}-\frac{\partial^2 w}{\partial x^2}=0$, for $t>0$ and two initial conditions

$$w=\varphi_0(x)\ \ at\ \ t=0,$$

$$\frac{\partial w}{\partial t}=\varphi_1(x)\ at\ \ t=0.$$

The solution of the Cauchy problem is given by D'Alembert's formula:

$$w(x,t)=\frac{1}{2}[\varphi_0(x+at)+\varphi_0(x-at)]+\frac{1}{2a}\int_{x-at}^{x+at}\varphi_1(\xi)d\xi.$$

Boundary value problems. In all boundary value problems, it is required to find a function w, in a domain $t\geq 0,\ x_1\leq x\leq x_2\ (-\infty<x_1<x_2<\infty)$, that satisfies the wave equation $\frac{\partial^2 w}{\partial t^2}-\frac{\partial^2 w}{\partial x^2}=0,$ for $t>0$ and the initial conditions. In addition, appropriate boundary conditions are imposed.

Solutions to these boundary value problems for the wave equation can be obtained by separation of variables (Fourier method) in the form of infinite series. In particular, the solution to the first boundary value problem with homogeneous boundary conditions, $\psi_1(t)=\psi_2(t)\equiv 0\ at\ x_1=0\ and\ x_2=l$, is expressed as,

$$w(x,t)=\frac{\partial}{\partial t}\int_0^l\varphi_0(\xi)G(x,\xi,t)d\xi+\int_0^l\varphi_1(\xi)G(x,\xi,t)d\xi,$$

Where

$$G(x,\xi,t)=\frac{2}{a\pi}\sum_{n=1}^{\infty}\frac{1}{n}sin\left(\frac{n\pi x}{l}\right)sin\left(\frac{n\pi\xi}{l}\right)sin\left(\frac{n\pi at}{l}\right).$$

Goursat problem:

On the characteristics of the wave equation $\frac{\partial^2 w}{\partial t^2}-\frac{\partial^2 w}{\partial x^2}=0$, values of the unknown function w are prescribed:

$$\begin{aligned}&w=\varphi(x)\quad for\quad x-t=0\ (0\leq x\leq a),\\&w=\psi(x)\quad for\quad x+t=0\ \ (0\leq x\leq b),\end{aligned}$$

with the consistency condition $\varphi(0)=\psi(0)$ implied to hold.

Substituting the values set on the characteristics into the general solution of the wave equation $w=\varphi(x+t)+\psi(x-t)$, one arrives at a system of linear algebraic equations for $\varphi(x)$ and $\psi(x)$. As a result, the solution to the Goursat problem is obtained in the form

$$w(x,t)=\varphi\left(\frac{x+t}{2}\right)+\psi\left(\frac{x-t}{2}\right)-\varphi(0).$$

The solution propagation domain is the parallelogram bounded by the four lines

$$x-t=0,\quad x+t=0,\quad x-t=2b,\quad x+t=2a.$$

Boundary Value Problems for Elliptic Equations

Setting boundary conditions for the first, second, and third boundary value problems for the Laplace equation $\frac{\partial^2 w}{\partial x^2}+\frac{\partial^2 w}{\partial y^2}=0$, means prescribing values of the unknown function, its first derivative, and a linear combination of the unknown function and its derivative, respectively.

For example, the first boundary value problem in a rectangular domain $0 \le x \le a, 0 \le y \le b$ is characterized by the boundary conditions

$$w=\varphi_1(y) \;\; at \;\; x=0, \quad w=\varphi_2(y) \; at \;\; x=a,$$
$$w=\varphi_3(x) \;\; at \;\; y=0, \quad w=\varphi_4(x) \; at \;\; y=b.$$

The solution to problem,with the $\varphi_3(x)=\varphi_4(x)\equiv 0$ is given by

$$w(x,y)=\sum_{n=1}^{\infty} A_n \, sinh\left[\frac{n\pi}{b}(a-x)\right]sin\left(\frac{n\pi}{b}y\right)+\sum_{n=1}^{\infty} B_n \, sinh\left(\frac{n\pi}{b}x\right)sin\left(\frac{n\pi}{b}y\right),$$

Where the coefficients A_n and B_n are expressed as

$$A_n=\frac{2}{\lambda_n}\int_0^b \varphi_1(\xi) sin\left(\frac{n\pi\xi}{b}\right)d\xi, \;\; B_n=\frac{2}{\lambda n}\int_0^b \varphi_2(\xi) sin\left(\frac{n\pi\xi}{b}\right)d\xi, \;\; \lambda_n=b \, sinh\left(\frac{n\pi a}{b}\right).$$

For elliptic equations, the first boundary value problem is often called the Dirichlet problem, and the second boundary value problem is called the Neumann problem.

Some Nonlinear Equations Encountered in Applications

Nonlinear heat equation:

$$\frac{\partial w}{\partial t}=\frac{\partial}{\partial x}\left[f(w)\frac{\partial w}{\partial x}\right].$$

This equation describes one-dimensional unsteady thermal processes in quiescent media or solids in the case where the thermal diffusivity is temperature dependent, $f(w)>0$. In the special case $f(w)\equiv 1$, the nonlinear equation $\frac{\partial w}{\partial t}=\frac{\partial}{\partial x}\left[f(w)\frac{\partial w}{\partial x}\right]$ becomes the linear heat equation $\frac{\partial w}{\partial t}-\frac{\partial^2 w}{\partial x^2}=0$.

In general, the nonlinear heat equation $\frac{\partial w}{\partial t}=\frac{\partial}{\partial x}\left[f(w)\frac{\partial w}{\partial x}\right]$ admits exact solutions of the form,

$w=W(kx-\lambda t)$ (traveling-wave solution)

$w = U(x/\sqrt{t})$, (self-similar solution),

where $W = W(z)$ and $U = U(r)$ are determined by ordinary differential equations, and k and λ are arbitrary constants.

Kolmogorov–Petrovskii–Piskunov equation:

$$\frac{\partial w}{\partial t} = a\frac{\partial^2 w}{\partial x^2} + f(w), \qquad a > 0.$$

Equations of this form are often encountered in various problems of mass and heat transfer (with f being the rate of a volume chemical reaction), combustion theory, biology, and ecology.

In the special case of $f(w) \equiv 0$ and $a = 1$, the nonlinear equation $\frac{\partial w}{\partial t} = a\frac{\partial^2 w}{\partial x^2} + f(w)$, $a > 0$ becomes the linear heat equation $\frac{\partial w}{\partial t} - \frac{\partial^2 w}{\partial x^2} = 0$.

Remark. Equation $\frac{\partial w}{\partial t} = a\frac{\partial^2 w}{\partial x^2} + f(w)$, $a > 0$ is also called a heat equation with a nonlinear source.

Burgers equation:

$$\frac{\partial w}{\partial t} + w\frac{\partial w}{\partial x} = \frac{\partial^2 w}{\partial x^2}.$$

This equation is used for describing wave processes in gas dynamics, hydrodynamics, and acoustics.

1. Exact solutions to the Burgers equation can be obtained using the following formula (Hopf–Cole transformation):

$$w(x,t) = -\frac{2}{u}\frac{\partial u}{\partial x},$$

where $u=u(x,t)$ is a solution to the linear heat equation $u_t = u_{xx}$.

2. The solution to the Cauchy problem for the Burgers equation with the initial condition

$$w = f(x) \quad \text{at} \quad t = \quad (-\infty < x < \infty)$$

has the form

$$w(x,t) = -2\frac{\partial}{\partial x}\ln F(x,t),$$

where,

$$F(x,t) = \frac{1}{\sqrt{4\pi t}}\int_{-\infty}^{\infty}\exp\left[-\frac{(x-\xi)^2}{4t} + \frac{1}{2}\int_0^{\xi} f(\xi')\,d\xi'\right]d\xi$$

Nonlinear wave equation:

$$\frac{\partial^2 w}{\partial t^2} = \frac{\partial}{\partial x}\left[f(w)\frac{\partial w}{\partial x}\right].$$

This equation is encountered in wave and gas dynamics, $f(w)>0$. In the special case $f(w)\equiv 1$, the nonlinear equation $\frac{\partial^2 w}{\partial t^2} = \frac{\partial}{\partial x}\left[f(w)\frac{\partial w}{\partial x}\right]$. becomes the linear wave equation $\frac{\partial^2 w}{\partial t^2} - \frac{\partial^2 w}{\partial x^2} = 0.$

Equation $\frac{\partial^2 w}{\partial t^2} = \frac{\partial}{\partial x}\left[f(w)\frac{\partial w}{\partial x}\right]$. admits exact solutions in implicit form:

$$x + t\sqrt{f(w)} = \varphi(w),$$
$$x - t\sqrt{f(w)} = \psi(w),$$

where $\varphi(w)$ and $\psi(w)$ are arbitrary functions.

Equation $\frac{\partial^2 w}{\partial t^2} = \frac{\partial}{\partial x}\left[f(w)\frac{\partial w}{\partial x}\right]$. can be reduced to a linear equation.

Nonlinear Klein–Gordon Equation

$$\frac{\partial_2 w}{\partial t^2} = a\frac{\partial^2 w}{\partial x^2} + f(w), \quad a>0.$$

Equations of this form arise in differential geometry and various areas of physics (superconductivity, dislocations in crystals, waves in ferromagnetic materials, laser pulses in two-phase media, and others). For $f(w)\equiv 0$ and $a=1$, equation $\frac{\partial_2 w}{\partial t^2} = a\frac{\partial^2 w}{\partial x^2} + f(w),\ a>0.$ coincides with the linear wave equation $\frac{\partial^2 w}{\partial t^2} - \frac{\partial^2 w}{\partial x^2} = 0.$

1. In general, the nonlinear Klein–Gordon equation $\frac{\partial_2 w}{\partial t^2} = a\frac{\partial^2 w}{\partial x^2} + f(w), \quad a>0.$ admits exact solutions of the form

$$w = W(z),\ z = kx - \lambda t$$
$$w = U(\xi),\ ,\xi = (\sqrt{a}t + C_1) - (x + C_2)^2,$$

where $W=W(z)$ and $U=U(\xi)$ are determined by ordinary differential equations, while k, λ, C_1, and C_2 are arbitrary constants.

2. In the special case

$$f(w) = be^{\beta w},$$

the general solution of equation $\frac{\partial_2 w}{\partial t^2} = a\frac{\partial^2 w}{\partial x^2} + f(w), \quad a > 0.$ is expressed as

$$w(x,t) = \frac{1}{\beta}[\varphi(z) + \psi(y)] - \frac{2}{\beta}\ln\left|k\int \exp[\varphi(z)]dz - \frac{b\beta}{8ak}\int \exp[\psi(y)]dy\right|,$$

$$z = x - \sqrt{at}, \quad y = x + \sqrt{at},$$

where $\varphi=\varphi(z)$ and $\psi=\psi(y)$ are arbitrary functions and k is an arbitrary constant.

In the special cases $f(w)=b\sin(\beta w)$ and $f(w)=b\sinh(\beta w)$, equation $\frac{\partial_2 w}{\partial t^2} = a\frac{\partial^2 w}{\partial x^2} + f(w), \quad a > 0.$ is called the sine-Gordon equation and the sinh-Gordon equation, respectively.

Nonlinear Laplace Equation

$$\frac{\partial^2 w}{\partial x^2} + \frac{\partial^2 w}{\partial y^2} = f(w).$$

This equation is also called a *stationary heat equation with a nonlinear source.*

1. In general, the nonlinear heat equation $\frac{\partial^2 w}{\partial x^2} + \frac{\partial^2 w}{\partial y^2} = f(w).$ admits exact solutions of the form

$$w = W(z), \qquad z = k_1 x + k_2 y,$$

$$w = U(r), \qquad r = \sqrt{(x + C_1)^2 + (y + C_2)^2},$$

where $W=W(z)$ and $U=U(r)$ are determined by ordinary differential equations, while k_1, k_2, C_1, and C_2 are arbitrary constants.

2. In the special case

$$f(w) = ae^{\beta w},$$

the general solution of equation (32) is expressed as

$$w(x,y) = -\frac{2}{\beta}\ln\frac{|1 - 2a\beta\Phi(z)\overline{\Phi(z)}|}{4|\Phi' z(z)|},$$

where $\Phi = \Phi(z)$ is an arbitrary analytic function of the complex variable $z = x + iy$ with nonzero derivative, and the bar over a symbol denotes the complex conjugate.

Monge–Ampere Equation

$$\left(\frac{\partial^2 w}{\partial x \partial y}\right)^2 - \frac{\partial^2 w \partial^2 w}{\partial x^2\, \partial y^2} = f(x,y).$$

The equation is encountered in differential geometry, gas dynamics, and meteorology.

Below are solutions to the homogeneous Monge–Ampere equation for the special case $f(x,y) \equiv 0$.

1. Exact solutions involving one arbitrary function:

$$w(x,y) = \varphi(C_{1x} + C_{2y}) + C_{3x} + C_{4y} + C_5,$$

$$w(x,y) = (C_{1x} + C_{2y})\varphi\left(\frac{y}{x}\right) + C_{3x} + C_{4y} + C_5,$$

$$w(x,y) = (C_{1x} + C_{2y} + C_3)\varphi\left(\frac{C_{4x} + C_{5y} + C_6}{C_{1x} + C_{2y} + C_3}\right) + C_{7x} + C_8 y + C_9,$$

where C_1, ..., C_9 are arbitrary constants and $\varphi = \varphi(z)$ is an arbitrary function.

2. General solution in parametric form:

$$w = tx + \varphi(t)y + \psi(t),$$
$$x + \varphi'(t)y + \psi'(t) = 0,$$

where t is the parameter, and $\varphi = \varphi(t)$ and $\psi = \psi(t)$ are arbitrary functions.

Simplest Types of Exact Solutions of Nonlinear PDEs

Preliminary remarks

The following classes of solutions are usually regarded as exact solutions to nonlinear partial differential equations of mathematical physics:

1. Solutions expressible in terms of elementary functions.
2. Solutions expressed by quadrature.
3. Solutions described by ordinary differential equations (or systems of ordinary differential equations).
4. Solutions expressible in terms of solutions to linear partial differential equations (and solutions to linear integral equations).

The simplest types of exact solutions to nonlinear PDEs are traveling-wave solutions and self-similar solutions. They often occur in various applications.

In what follows, it is assumed that the unknown w depends on two variables, x and t, where $t|$ plays the role of time and x is a spatial coordinate.

Traveling-wave Solutions

Traveling-wave solutions, by definition, are of the form

$$w(x,t) = W(z), \qquad z = kx - \lambda t,$$

Where λ / k plays the role of the wave propagation velocity (the value $\lambda = 0$ corresponds to a stationary solution, and the value $k = 0$ corresponds to a space-homogeneous solution). Traveling-wave solutions are characterized by the fact that the profiles of these solutions at different time instants are obtained from one another by appropriate shifts (translations) along the x-axis. Consequently, a Cartesian coordinate system moving with a constant speed can be introduced in which the profile of the desired quantity is stationary. For $k > 0$ and $\lambda > 0$, the wave travels along the x-axis to the right (in the direction of increasing x).

Traveling-wave solutions occur for equations that do not explicitly involve independent variables,

$$F\left(w, \frac{\partial w}{\partial x}, \frac{\partial w}{\partial t}, \frac{\partial^2 w}{\partial x^2}, \frac{\partial^2 w}{\partial x \partial t}, \frac{\partial^2 w}{\partial t^2}, \ldots\right) = 0.$$

Substituting into one obtains an autonomous ordinary differential equation for the function $W(z)$:

$$F(W, kW', -\lambda W', k^2 W'', -k\lambda W'', \lambda^2 W'', \ldots) = 0,$$

where k and λ are arbitrary constants, and the prime denotes a derivative with respect to z.

The term traveling-wave solution is also used in the cases where the variable t plays the role of a spatial coordinate, $t = y$.

All nonlinear equations considered above, and with $f(x, y) = 0$, admit traveling-wave solutions.

Self-similar solutions

By definition, a self-similar solution is a solution of the form

$$w(x,t) = t^{\alpha} U(\zeta), \qquad \zeta = x t^{\beta}.$$

The profiles of these solutions at different time instants are obtained from one another by a similarity transformation (like scaling).

Self-similar solutions exist if the scaling of the independent and dependent variables,

$$t = C\bar{t}, x = C^k \bar{x}, w = C^m \bar{w}, \quad \text{where } C \neq 0 \text{ is an arbitrary constant,}$$

for some k and m such that $|k|+|m|\neq 0$, is equivalent to the identical transformation.

It can be shown that the parameters in solution and transformation are linked by the simple relations

$$\alpha = m, \qquad \beta = -k.$$

In practice, the above existence criterion is checked and if a pair of k and m in has been found, then a self-similar solution is defined by formulas with parameters.

Example: Consider the heat equation with a nonlinear power-law source term

$$\frac{\partial w}{\partial t} = a\frac{\partial^2 w}{\partial x^2} + bw^n$$

The scaling transformation converts equation $\frac{\partial w}{\partial t} = a\frac{\partial^2 w}{\partial x^2} + bw^n$ into

$$C^{m-1}\frac{\partial \overline{w}}{\partial t} = aC^{m-2k}\frac{\partial 2\overline{w}}{\partial x2} + bC^{mn}\overline{w}^n.$$

In order that equation $C^{m-1}\frac{\partial \overline{w}}{\partial t} = aC^{m-2k}\frac{\partial 2\overline{w}}{\partial x2} + bC^{mn}\overline{w}^n$. coincides with one must require that the powers of C are the same, which yields the following system of linear algebraic equations for the constants k and m:

$$m-1 = m-2k = mn.$$

This system admits a unique solution

$$k = \frac{1}{2},$$

$m = \frac{1}{1-n}$ .Using this solution together with relations and, one obtains self-similar variables in the form

$$w = t^{1/(1-n)}U(\zeta), \quad \zeta = xt^{-1/2}.$$

Inserting these into, one arrives at the following ordinary differential equation for $U(\zeta)$:

$$aU''_{\zeta\zeta} + \frac{1}{2}\zeta U'_{\zeta} + \frac{1}{n-1}U + bU^n = 0.$$

Cauchy Problem and Boundary Value Problems for Nonlinear Equations

The Cauchy problem and boundary value problems for nonlinear equations are stated in exactly the same way as for linear equations.

Examples: The Cauchy problem for a nonlinear heat equation is stated as follows: find a solution to equation $\frac{\partial w}{\partial t} = \frac{\partial}{\partial x}\left[f(w)\frac{\partial w}{\partial x} \right]$ subject to the initial condition.
The first boundary value problem for a nonlinear wave equation as follows: find a solution to equation $\frac{\partial^2 w}{\partial x^2} + \frac{\partial^2 w}{\partial y^2} = f(w)$. subject to the initial conditions and the boundary conditions.

Problems for nonlinear PDEs are normally solved using numerical methods.

Higher-Order Partial Differential Equations

Apart from second-order PDEs, higher-order equations also quite often arise in applications. Below are only a few important examples of such equations with some of their solutions.

Higher-Order Linear Partial Differential Equations

Equation of transverse vibration of elastic rod:

$$\frac{\partial^2 w}{\partial t^2} + a^2 \frac{\partial^4 w}{\partial x^4} = 0.$$

The equation has the following particular solutions:

$$w(x,t) = [A\sin(\lambda x) + B\cos(\lambda x) + C\sin h(\lambda x) + D\cos(\lambda x)]\sin(\lambda^2 at),$$
$$w(x,t) = [A_1\sin(\lambda x) + B_1\cos(\lambda x) + C_1\sin h(\lambda x) + D_1\cos(\lambda x)]\cos(\lambda^2 at)$$

where A, B, C, D , A_1, B_1, C_1, D_1, and λ are arbitrary constants.

Biharmonic Equation

$$\Delta\Delta w = 0$$

where $\Delta\Delta$ is the biharmonic operator,

$$\Delta\Delta \equiv \Delta^2 = \frac{\partial^4}{\partial x^4} + 2\frac{\partial^4}{\partial x^2 \partial y^2} + \frac{\partial^4}{\partial y^4}.$$

The biharmonic equation $\Delta\Delta w = 0$ is encountered in plane problems of elasticity (wis the Airy stress function). It is also used to describe slow flows of viscous incompressible fluids (w is the stream function).

Various representations of the general solution to equation $\Delta\Delta w = 0$ in terms of harmonic functions include,

$$w(x,y) = xu_1(x,y) + u_2(x,y),$$
$$w(x,y) = yu_1(x,y) + u_2(x,y),$$
$$w(x,y) = (x^2 + y^2)u_1(x,y) + u_2(x,y),$$

where u_1 and u_2 are arbitrary functions satisfying the Laplace equation $\Delta u_k = 0 (k = 1, 2)$.

Complex form of representation of the general solution:

$$w(x,y) = \mathrm{Re}\left[\bar{z}\, f(z) + g(z)\right],$$

where $f(z)$ and $g(z)$ are arbitrary analytic functions of the complex variable $z = x + iy$; $\bar{z} = x - iy, i^2 = -1$. The symbol $\mathrm{Re}[A]$ stands for the real part of a complex quantity A.

Higher-Order Nonlinear Partial Differential Equations

Korteweg–de Vries equation:

$$\frac{\partial w}{\partial t}\frac{\partial^3 w}{\partial x^3} - 6w\frac{\partial w}{\partial x} = 0.$$

Equation of a steady laminar boundary layer on a flat plate:

$$\frac{\partial w}{\partial y}\frac{\partial^2 w}{\partial x \partial y} - \frac{\partial w}{\partial x}\frac{\partial^2 w}{\partial y^2} = a\frac{\partial^3 w}{\partial y^3},$$

where w is the stream function.

Boussinesq equation:

$$\frac{\partial^2 w}{\partial t^2} + \frac{\partial}{\partial x}\left(w\frac{\partial w}{\partial x}\right) + \frac{\partial^4 w}{\partial x^4} = 0.$$

This equation arises in several physical applications: propagation of long waves in shallow water, one-dimensional nonlinear lattice-waves, vibrations in a nonlinear string, and ion sound waves in a plasma.

Equation of motion of a viscous fluid:

$$\frac{\partial w}{\partial y}\frac{\partial}{\partial x}(\Delta w)-\frac{\partial w}{\partial x}\frac{\partial}{\partial y}(\Delta w)=a\Delta\Delta w, \qquad \Delta w=\frac{\partial^2 w}{\partial x^2}+\frac{\partial^2 w}{\partial y^2}$$

This is a two-dimensional stationary equation of motion of a viscous incompressible fluid—it is obtained from the Navier–Stokes equations by the introduction of the stream function.

Approximate and Numerical Methods

The preceding discussion pertains to the exact or analytical solution of PDEs. For example, in the case of a heat equation or a wave equation, an exact solution would be a function $w=f(x,t)$ which, when substituted into the respective equation would satisfy it identically along with all of the associated initial and boundary conditions.

Although analytical solutions are exact, they also may not be available, simply because we do not know how to derive such solutions. This could be because the PDE system has too many PDEs, or they are too complicated, e.g., nonlinear, or both, to be amenable to analytical solution. In this case, we may have to resort to an approximate solution. That is, we seek an analytical or numerical approximation to the exact solution.

Perturbation methods are an important subset of approximate analytical methods. They may be applied if the problem involves small (or large) parameters, which are used for constructing solutions in the form of asymptotic expansions.

Unlike exact and approximate analytical methods, methods to compute numerical PDE solutions are in principle not limited by the number or complexity of the PDEs. This generality combined with the availability of high performance computers makes the calculation of numerical solutions feasible for a broad spectrum of PDEs (such as the Navier–Stokes equations) that are beyond analysis by analytical methods. The development and implementation (as computer codes) of numerical methods or algorithms for PDE systems is a very active area of research. Here we indicate in the external links just two readily available links to Scholarpedia.

Parabolic PDE

Analytical solutions to a parabolic PDE (heat equation) are given here. But we will proceed with a numerical solution and use one of these analytical solutions to evaluate the numerical solution.

We can consider the numerical solution to the heat equation,

$$\frac{\partial w}{\partial t}-\frac{\partial^2 w}{\partial x^2}=0$$

As a two-step process:

1. Numerical approximation of the derivative $\frac{\partial^2 w}{\partial x^2}$. At this point, we will have a semi-discretization of Eq. $\frac{\partial w}{\partial t} - \frac{\partial^2 w}{\partial x^2} = 0$.
2. Numerical approximation of the derivative $\frac{\partial w}{\partial t}$. At this point, we will have a full discretization of Eq. $\frac{\partial w}{\partial t} - \frac{\partial^2 w}{\partial x^2} = 0$.

In order to implement these two steps, we require a grid in x and a grid in t. For the grid in x, we denote a position along the grid with the index i. Then we can consider the Taylor series expansion of the numerical solution at grid point i,

$$w_{i+1} = w_i + \frac{dw_i}{dx}(x_{i+1} - x_i) + \frac{d^2w_i}{dx^2}\frac{(x_{i+1} - x_i)^2}{2!} + \frac{d^2w_i}{dx^3}\frac{(x_{i+1} - x_i)^3}{3!} +$$

and

$$w_{i-1} = w_i + \frac{dw_i}{dx}(x_{i-1} - x_i) + \frac{d^2w_i}{dx^2}\frac{(x_{i-1} - x_i)^2}{2!} + \frac{d^2w_i}{dx^3}\frac{(x_{i-1} - x_i)^3}{3!} + \ldots$$

If we consider a uniform grid (a grid with uniform spacing $\Delta x = x_{i+1} - x_i = x_i - x_{i-1}$), addition of Eqs. $w_{i+1} = w_i + \frac{dw_i}{dx}(x_{i+1} - x_i) + \frac{d^2w_i}{dx^2}\frac{(x_{i+1} - x_i)^2}{2!} + \frac{d^2w_i}{dx^3}\frac{(x_{i+1} - x_i)^3}{3!} + \ldots$ and $w_{i-1} = w_i + \frac{dw_i}{dx}(x_{i-1} - x_i) + \frac{d^2w_i}{dx^2}\frac{(x_{i-1} - x_i)^2}{2!} + \frac{d^2w_i}{dx^3}\frac{(x_{i-1} - x_i)^3}{3!} + \ldots$ gives (note the cancellation of the first and third derivative terms since $x_{i-1} - x_i = -\Delta x$),

$$w_{i+1} + w_{i-1} = 2w_i + \frac{d^2w_i}{dx^2}\Delta x^2 + O(\Delta x^4)$$

where $O\left(\Delta x^4\right)$ denotes a term proportional to Δx^4 or of order Δx^4; this term can be considered a truncation error resulting from truncating the Taylor series of Eqs. $w_{i+1} = w_i + \frac{dw_i}{dx}(x_{i+1} - x_i) + \frac{d^2w_i}{dx^2}\frac{(x_{i+1} - x_i)^2}{2!} + \frac{d^2w_i}{dx^3}\frac{(x_{i+1} - x_i)^3}{3!} + \ldots$ and $w_{i-1} = w_i + \frac{dw_i}{dx}(x_{i-1} - x_i) + \frac{d^2w_i}{dx^2}\frac{(x_{i-1} - x_i)^2}{2!} + \frac{d^2w_i}{dx^3}\frac{(x_{i-1} - x_i)^3}{3!} + \ldots$ beyond the Δx^2 term. Then Eq. $w_{i+1} + w_{i-1} = 2w_i + \frac{d^2w_i}{dx^2}\Delta x^2 + O(\Delta x^4)$ gives for the second derivative $\frac{d^2w_i}{dx^2} \approx \frac{w_{i+1} - 2w_i + w_{i-1}}{\Delta x^2} + O(\Delta x^2)$.

Equation $\frac{d^2 w_i}{dx^2} \approx \frac{w_{i+1} - 2w_i + w_{i-1}}{\Delta x^2} + O(\Delta x^2)$ is a second order (because of the principal error or truncation error $O\left(\Delta x^2\right)$) finite difference approximation of $d^2 wi / dx^2$.

If Eq. $\frac{d^2 w_i}{dx^2} \approx \frac{w_{i+1} - 2w_i + w_{i-1}}{\Delta x^2} + O(\Delta x^2)$ is substituted in Eq. $\frac{\partial w}{\partial t} - \frac{\partial^2 w}{\partial x^2} = 0$ (to replace the derivative $\frac{\partial^2 w}{\partial x^2}$), a system of ODEs results,

$$\frac{dw_i}{dt} = D\frac{w_{i+1} - 2w_i + w_{i-1}}{\Delta x^2} + O(\Delta x^2), i = 1, 2, \ldots, N$$

(we have added a multiplying constant D to the right-hand side of Eq. $\frac{\partial w}{\partial t} - \frac{\partial^2 w}{\partial x^2} = 0$ generally termed a thermal diffusivity if w in Eq. $\frac{\partial w}{\partial t} - \frac{\partial^2 w}{\partial x^2} = 0$ is temperature and a mass diffusivity if w is concentration; D has the MKS units m^2/s as expected from a consideration of Eq. $\frac{\partial w}{\partial t} - \frac{\partial^2 w}{\partial x^2} = 0$ with x in metres and t in seconds).

Note that the independent variable x does not appear explicitly in Eqs. $\frac{dw_i}{dt} = D\frac{w_{i+1} - 2w_i + w_{i-1}}{\Delta x^2} + O(\Delta x^2), i = 1, 2, \ldots, N$ and that the only independent variable is t (so that they are ODEs). N is the number of points in the x grid (x is termed a boundary value variable since the terminal grid points at $i = 1$ and $i = N$ typically refer to the boundaries of a physical system). Thus Eq. $\frac{\partial w}{\partial t} - \frac{\partial^2 w}{\partial x^2} = 0$ is partly discretized (in x) and therefore Eqs. $\frac{dw_i}{dt} = D\frac{w_{i+1} - 2w_i + w_{i-1}}{\Delta x^2} + O(\Delta x^2), i = 1, 2, \ldots, N$ are referred to as a semi-discretization.

To compute a solution to Eq. $\frac{\partial w}{\partial t} - \frac{\partial^2 w}{\partial x^2} = 0$, we could apply an established initial-value integrator in t. This is the essence of the method of lines (MOL). Alternatively, we could now discretize Eqs. $\frac{dw_i}{dt} = D\frac{w_{i+1} - 2w_i + w_{i-1}}{\Delta x^2} + O(\Delta x^2), i = 1, 2, \ldots, N$. For example, if we apply Eq. $w_{i+1} = w_i + \frac{dw_i}{dx}(x_{i+1} - x_i) + \frac{d^2 w_i}{dx^2}\frac{(x_{i+1} - x_i)^2}{2!} + \frac{d^2 w_i}{dx^3}\frac{(x_{i+1} - x_i)^3}{3!} + \ldots$ on a grid in t with an index k,

$$w_i^{k+1} = w_i^k + \frac{dw_i^k}{dt}(t^{k+1} - t^k) + \frac{d^2 w_i^k}{dt^2}\frac{(t^{k+1} - t^k)^2}{2!} + \ldots, \quad i = 1, 2, \ldots, N, \quad k = 1, 2, \ldots$$

If the grid in t has a uniform spacing $h = t^{k+1} - t^k$ and if truncation after the first derivative term is applied,

$$w_i^{k+1} = w_i^k + \frac{dw_i^k}{dt}h + O(h^2)$$

Equation $w_i^{k+1} = w_i^k + \frac{dw_i^k}{dt}h + O(h^2)$, the classical Euler's method, can be used to step along the solution of Eq. $\frac{\partial w}{\partial t} - \frac{\partial^2 w}{\partial x^2} = 0$ from point k to $k+1$ (at a grid point i in x).

Application of Eq. $w_i^{k+1} = w_i^k + \frac{dw_i^k}{dt}h + O(h^2)$ to Eq. $\frac{dw_i}{dt} = D\frac{w_{i+1} - 2w_i + w_{i-1}}{\Delta x^2} + O(\Delta x^2), i = 1, 2, \ldots, N$ gives the fully discretized approximation of Eq. $\frac{\partial w}{\partial t} - \frac{\partial^2 w}{\partial x^2} = 0$,

$$w_i^{k+1} + w_i^k + hD\frac{w_{i+1}^k - 2w_i^k + w_{i-1}^k}{\Delta x^2}$$

In Eq. $w_i^{k+1} = w_i^k + \frac{dw_i^k}{dt}(t^{k+1} - t^k) + \frac{d^2 w_i^k}{dt^2}\frac{(t^{k+1} - t^k)^2}{2!} + \ldots, \quad i = 1, 2, \ldots, N, \quad k = 1, 2, \ldots$ we do not specify the total number of grid points in t (as we did with the grid in x); t is an initial value variable since it is typically time, and is defined over the semiinfinite interval $0 \le t \le \infty$.

Note that Eq. $w_i^{k+1} + w_i^k + hD\frac{w_{i+1}^k - 2w_i^k + w_{i-1}^k}{\Delta x^2}$ explicitly gives the solution at the advanced point in t $(at\, k+1)$ and therefore it is an explicit finite difference approximation of Eq. $\frac{\partial w}{\partial t} - \frac{\partial^2 w}{\partial x^2} = 0$. We can now consider using Eq. $w_i^{k+1} + w_i^k + hD\frac{w_{i+1}^k - 2w_i^k + w_{i-1}^k}{\Delta x^2}$ to step forward from an initial condition (IC) required by Eq. $\frac{\partial w}{\partial t} - \frac{\partial^2 w}{\partial x^2} = 0$. Here we take as the initial condition,

$$w(x, t = 0) = Ae^{-(\mu/D)x} + B$$

where A, B, μ are constants to be specified. The finite difference form of Eq. $w(x, t = 0) = Ae^{-(\mu/D)x} + B$ is,

$$w_i^0 = Ae^{-(\mu/D)x_i} + B.$$

(note that $k=0 \quad at \quad t=0$)

We must also specify two boundary conditions (BCs) for Eq. $\frac{\partial w}{\partial t}-\frac{\partial^2 w}{\partial x^2}=0$ (since it is second order in x). We will use the Dirichlet BC at $x=0$

$$w(x=0,t)=Ae^{-(\mu/D)(-\mu t)}+B.$$

for which the finite difference form is (note that $i=1 \quad at \quad x=0$)

$$w_1^k=Ae^{-(\mu/D)(-\mu t^k)}+B.$$

We use the Neumann BC at $x=1$

$$\frac{\partial w(x=1,t)}{\partial x}=A(-\mu/D)e^{-(\mu/D)(1-\mu t)}.$$

for which the finite difference form is (note that $i=N \quad at \quad x=1$)

$$w_{N+1}^k=w_{N-1}^k+2\Delta xAe^{-(\mu/D)(-\mu t^k)}+B.$$

where w_{N+1} is a fictitious value that is outside the interval $0\le x\le 1$; it can be used to eliminate w N in Eq. $w_i^{k+1}+w_i^k+hD\frac{w_{i+1}^k-2w_i^k+w_{i-1}^k}{\Delta x^2}$ for $i=N$.

Equations $w_i^{k+1}+w_i^k+hD\frac{w_{i+1}^k-2w_i^k+w_{i-1}^k}{\Delta x^2}$, $w_i^0=Ae^{-(\mu/D)x_i}+B$, $w_1^k=Ae^{-(\mu/D)(-\mu t^k)}+B$ and $w_{N+1}^k=w_{N-1}^k+2\Delta xAe^{-(\mu/D)(-\mu t^k)}+B$ constitute the full system of equations for the calculation of the numerical solution to Eq. $\frac{\partial w}{\partial t}-\frac{\partial^2 w}{\partial x^2}=0$. Note that we have replaced the original PDE, Eq. $\frac{\partial w}{\partial t}-\frac{\partial^2 w}{\partial x^2}=0$, with a set of approximating algebraic equations (Eqs. $w_i^{k+1}+w_i^k+hD\frac{w_{i+1}^k-2w_i^k+w_{i-1}^k}{\Delta x^2}$, $w_i^0=Ae^{-(\mu/D)x_i}+B$, $w_1^k=Ae^{-(\mu/D)(-\mu t^k)}+B$ and $w_{N+1}^k=w_{N-1}^k+2\Delta xAe^{-(\mu/D)(-\mu t^k)}+B$ which can easily be programmed for a computer. Also, an analytical solution to Eq. $\frac{\partial w}{\partial t}-\frac{\partial^2 w}{\partial x^2}=0$ (see particular solutions to the heat equation) can be used to evaluate the numerical solution

$$w(x,t)=Ae^{-(1/D)(\mu^2 t\pm\mu x)}+B.$$

Equation above can be stated in the alternative form

$$w(x,t)=Ae^{-(\mu/D)(x-\mu t)}+B.$$

which corresponds to a traveling wave solution since x and t appear in the combination $x-\mu t$.

A short MATLAB program is based on Eqs. $w_i^{k+1}+w_i^k+hD\dfrac{w_{i+1}^k-2w_i^k+w_{i-1}^k}{\Delta x^2}$, $w_i^0=Ae^{-(\mu/D)x_i}+B$, $w_1^k=Ae^{-(\mu/D)(-\mu t^k)}+B$ and $w_{N+1}^k=w_{N-1}^k+2\Delta xAe^{-(\mu/D)(-\mu t^k)}+B$. Representative output from this program that compares the numerical solution from Eqs. $w_i^{k+1}+w_i^k+hD\dfrac{w_{i+1}^k-2w_i^k+w_{i-1}^k}{\Delta x^2}$, $w_i^0=Ae^{-(\mu/D)x_i}+B$, $w_1^k=Ae^{-(\mu/D)(-\mu t^k)}+B$ and $w_{N+1}^k=w_{N-1}^k+2\Delta xAe^{-(\mu/D)(-\mu t^k)}+B$ with the analytical solution, Eq. $w(x,t)=Ae^{-(\mu/D)(x-\mu t)}+B$, indicates that the two solutions are in agreement to five figures, as reflected in table.

Table: Comparison of the numerical and analytical solutions $x=x_l/^2=0.5$ produced by the program; w, numerical solution from Eqs. $w_i^{k+1}+w_i^k+hD\dfrac{w_{i+1}^k-2w_i^k+w_{i-1}^k}{\Delta x^2}$, $w_1^k=Ae^{-(\mu/D)(-\mu t^k)}+B$, $w_1^k=Ae^{-(\mu/D)(-\mu t^k)}+B$ and $w_{N+1}^k=w_{N-1}^k+2\Delta xAe^{-(\mu/D)(-\mu t^k)}+B$; wa, analytical solution of Eq. $w(x,t)=Ae^{-(\mu/D)(x-\mu t)}+B$

t = 0.05	w(x=xl/2,t) = 1.6376	wa(x=xl/2,t) = 1.6376
t = 0.10	w(x=xl/2,t) = 1.6703	wa(x=xl/2,t) = 1.6703
t = 0.15	w(x=xl/2,t) = 1.7047	wa(x=xl/2,t) = 1.7047
t = 0.20	w(x=xl/2,t) = 1.7408	wa(x=xl/2,t) = 1.7408
t = 0.25	w(x=xl/2,t) = 1.7788	wa(x=xl/2,t) = 1.7788
t = 0.30	w(x=xl/2,t) = 1.8187	wa(x=xl/2,t) = 1.8187
t = 0.35	w(x=xl/2,t) = 1.8607	wa(x=xl/2,t) = 1.8607
t = 0.40	w(x=xl/2,t) = 1.9048	wa(x=xl/2,t) = 1.9048
t = 0.45	w(x=xl/2,t) = 1.9512	wa(x=xl/2,t) = 1.9512
t = 0.50	w(x=xl/2,t) = 2.0000	wa(x=xl/2,t) = 2.0000

The parameters that produced the numerical output in above table are listed in below table.

Table: Numerical values of parameters

Parameter	Description	Value
D	diffusivity in Eq. $\dfrac{\partial w}{\partial t}-\dfrac{\partial^2 w}{\partial x^2}=0$	1
μ, A, B	constants in Eqs. $w(x,t=0)=Ae^{-(\mu/D)x}+B$ - $w(x,t)=Ae^{-(\mu/D)(x-\mu t)}+B$	1, 1, 1

N, xl	number of grid points and length in x	21, 1
h, tf	grid spacing in t, final value of t	0.001, 0.5

Additional parameters follow from the values in table. Thus, the grid spacing in x is $\Delta x = 1/(21-1) = 0.05$. The number of steps in t taken along the solution is $0.5/0.001 = 500$.

Some of the parameters, particularly N and h, were determined by trial and error to achieve a numerical solution of acceptable accuracy (e.g., five significant figures). We can note two additional points about these values:

- Acceptable values of N and h could be determined by observing the errors in the numerical solution through comparison with the exact solution (Eq. $w(x,t) = Ae^{-(\mu/D)(x-\mu t)} + B$) as illustrated in table. However, in most PDE applications, an analytical solution is not available for assessing the accuracy of the numerical solution, and in fact, the motivation for using a numerical method is generally to produce a solution when an analytical solution is not available. In this case (no analytical solution), a useful procedure for estimating the numerical accuracy is to compute solutions for two different values of N and compare the numerical values. If the two solutions do not agree to an acceptable level, a third solution is computed with a still larger N (smaller grid spacing in x) and again the solutions are compared. Eventually, if spatial convergence is achieved, successive solutions will agree to the required accuracy. In this case, the accuracy of the numerical solution is inferred, but the exact error is not computed (or even known) when an analytical solution is not available. The same reasoning can be applied for temporal convergence with respect to t, i.e., h is reduced until the successive solutions agree to a specified level.
- The preceding discussion was directed to achieving acceptable accuracy. Additionally, the values of N and h were selected to achieve a stable solution. The criterion for stability in the case of Eq. $w_i^{k+1} + w_i^k + hD\dfrac{w_{i+1}^k - 2w_i^k + w_{i-1}^k}{\Delta x^2}$ (for parabolic PDE (41)) is $\alpha = D\Delta t/\Delta x^2 < 1/2$. For the solution of table, $\alpha = 1\times 0.001/0.05^2 = 0.4 < 0.5$. If the critical value of the dimensionless parameter α=0.5 had been exceeded, the numerical solution would have become unstable (as manifest in numerical values of ever increasing magnitude). The stability constraint $\alpha < 0.5$ is a distinctive feature of the explicit finite difference approximation of Eq. $w_i^{k+1} + w_i^k + hD\dfrac{w_{i+1}^k - 2w_i^k + w_{i-1}^k}{\Delta x^2}$. Thus, as Δx is reduced (N increased) to achieve better accuracy in the numerical solution, h must also be reduced to maintain stability $(\alpha = Dh/\Delta x^2 < 0.5)$.

Hyperbolic PDE

We now consider a numerical solution to the one-dimensional hyperbolic wave equation

$$\frac{\partial^2 w}{\partial t^2}-\frac{\partial^2 w}{\partial x^2}=0.$$

We again have an analytical solution to evaluate the numerical solution. First, we include a velocity,

C, in the equation:

$$\frac{\partial^2 w}{\partial t^2}=c^2\frac{\partial^2 w}{\partial x^2}.$$

Note that c has the MKS units of m/s as expected and as inferred from Eq. $\frac{\partial^2 w}{\partial t^2}=c^2\frac{\partial^2 w}{\partial x^2}$ (note the units of the derivatives in x and t).

Since Eq. $\frac{\partial^2 w}{\partial t^2}$ $\frac{\partial^2 w}{\partial x^2}$ is second order in x and , it requires two ICs and two BCs. We will take these as:

$$w(x,t=0)=e^{-\lambda x^2}.$$

$$\frac{\partial w(x,t=0)}{\partial t}=0.$$

$$w(x\to\infty,t)=0.$$

$$w(x\to-\infty,t)=0$$

Equation $w(x,t=0)=e^{-\lambda x^2}$ indicates the IC is a Gaussian pulse with the positive constant λ to be specified. Equation $\frac{\partial w(x,t=0)}{\partial t}=0$ indicates $w(x,t)$ starts with zero "velocity". Equations $w(x\to\infty,t)=0-w(x\to-\infty,t)=0$ indicate that the solution $w(x,t)$ does not depart from the initial value of zero specified by IC. In other words, λ is chosen large enough that the IC is effectively zero at $x=\pm\infty$ and remains at this value for subsequent t.

An important difference between the parabolic problem of Eq. $\frac{\partial w}{\partial t}-\frac{\partial^2 w}{\partial x^2}=0$ and the hyperbolic problem of Eq. $\frac{\partial^2 w}{\partial t^2}=c^2\frac{\partial^2 w}{\partial x^2}$ is that the former is first order in t while the latter is second order in t. Therefore, in order to develop a numerical method for Eq.

$\frac{\partial^2 w}{\partial t^2} - \frac{\partial^2 w}{\partial x^2} = 0$ (or Eq. $\frac{\partial^2 w}{\partial t^2} = c^2 \frac{\partial^2 w}{\partial x^2}$), we need an algorithm that can accommodate second derivatives in t. While such algorithms do exist, they generally are not required. Rather, we can express a PDE second order in t as two PDEs first order in t. For example, Eq. $\frac{\partial^2 w}{\partial t^2} = c^2 \frac{\partial^2 w}{\partial x^2}$ can be written as

$$\frac{\partial w}{\partial t} = u .$$

$$\frac{\partial u}{\partial t} = c^2 \frac{\partial^2 w}{\partial x^2} .$$

Equations $\frac{\partial w}{\partial t} = u - \frac{\partial u}{\partial t} = c^2 \frac{\partial^2 w}{\partial x^2}$ are first order in t,and therefore an integration algorithm for first order equations, such as the Euler method of Eq. $w_i^{k+1} = w_i^k + \frac{dw_i^k}{dt} h + O(h^2)$, can be used to move $w(x,t)$ and $u(x,t)$ forward in t. Thus, the fully discretized form of Eqs. $\frac{\partial w}{\partial t} = u - \frac{\partial u}{\partial t} = c^2 \frac{\partial^2 w}{\partial x^2}$ can be written as

$$w_i^{k+1} = w_i^k + hu_i^k .$$

$$u_i^{k+1} = u_i^k + hc^2 \frac{w_{i+1}^k - 2w_i^k + w_{i-1}^k}{\Delta x^2} .$$

ICs become (with $k = 0$ corresponding to $t = 0$)

$$w_i^0 = e^{-\lambda x_i^2} .$$

$$u_i^0 = 0 .$$

For BCs, the infinite interval $-\infty \le x \le \infty$ must be replaced by a finite one $-x_l \le x \le x_l$ (since computers can accommodate only finite numbers) where xl is selected so that it is effectively infinite; that is, the solution $w(x,t)$ does not depart from IC at $x = -x_l, x_l$ for$t > 0$. The value x_l and the corresponding number of grid points in x N, are specified subsequently.

The finite difference approximations of BCs are

$$w_1^k = 0 .$$

$$w_N^k = 0 .$$

Equations $w_i^{k+1} = w_i^k + hu_i^k$, $u_i^{k+1} = u_i^k + hc^2 \frac{w_{i+1}^k - 2w_i^k + w_{i-1}^k}{\Delta x^2}$, $w_i^0 = e^{-\lambda x_i^2}$, $u_i^0 = 0$,

$w_1^k = 0$, $w_N^k = 0$ constitute the complete finite difference approximation of Eqs. $\frac{\partial w}{\partial t} = u$, $\frac{\partial u}{\partial t} = c^2 \frac{\partial^2 w}{\partial x^2}$, $w(x,t=0) = e^{-\lambda x^2}$, $\frac{\partial w(x,t=0)}{\partial t} = 0$, $w(x \to \infty, t) = 0$, $w(x \to -\infty, t) = 0$.

The general analytical solution to Eq. $\frac{\partial^2 w}{\partial t^2} = c^2 \frac{\partial^2 w}{\partial x^2}$ can be written as

$$w(x,t) = \frac{1}{2}[\varphi(x+ct) + \psi(x-ct)].$$

Let us take φ=ψ in the form of the Gaussian pulse of Eq. $w(x,t=0) = e^{-\lambda x^2}$, i.e.,

$$w(x,t) = \frac{1}{2}[e^{-\lambda(x+ct)^2} + e^{-\lambda(x-ct)^2}].$$

The plotted output from the program is given in figure and includes both the numerical solution of Eqs. $w_i^{k+1} = w_i^k + hu_i^k$, $u_i^{k+1} = u_i^k + hc^2 \frac{w_{i+1}^k - 2w_i^k + w_{i-1}^k}{\Delta x^2}$, $w_i^0 = e^{-\lambda x_i^2}$, $u_i^0 = 0$, $w_1^k = 0$, $w_N^k = 0$ and the analytical solution of Eq. $w(x,t) = \frac{1}{2}[e^{-\lambda(x+ct)^2} + e^{-\lambda(x-ct)^2}]$.

We can note the following points about figure below:

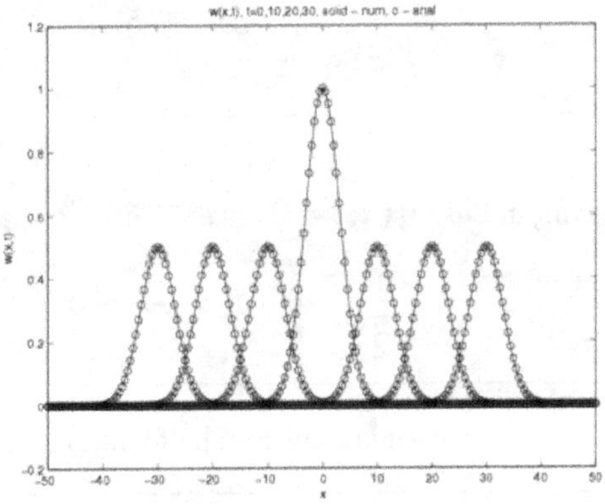

Figure: Comparison of the numerical and analytical solutions for t=0,10,20,30 produced by the program in Appendix 2; w, numerical solution from Eqs. $w_i^{k+1} = w_i^k + hu_i^k$, $u_i^{k+1} = u_i^k + hc^2 \frac{w_{i+1}^k - 2w_i^k + w_{i-1}^k}{\Delta x^2}$, $u_i^0 = 0$ $w_1^k = 0$, $w_N^k = 0$; wa, analytical solution of Eq. $w(x,t) = \frac{1}{2}[e^{-\lambda(x+ct)^2} + e^{-\lambda(x-ct)^2}]$

- The initial Gaussian pulse $t = 0$ (centered $x = 0$ with unit maximum value) splits into two pulses traveling left and right with velocity c=1 and maximum value of 0.5 according to Eq. $w(x,t) = \frac{1}{2}[e^{-\lambda(x+ct)^2} + e^{-\lambda(x-ct)^2}]$.
- The pulses traveling left are centered at $x = -10, -20, -30$. corresponding to $t = 10, 20, 30$ since $c = 1$. The pulses traveling right are centered at $x = 10, 20, 30$

corresponding $t = 10, 20, 30$. These properties are characteristic of the traveling wave functions of Eq. $w(x,t) = \frac{1}{2}[e^{-\lambda(x+ct)^2} + e^{-\lambda(x-ct)^2}]$ with arguments $x + ct$ and $x - ct$, respectively. In fact, the use of the word characteristic is particularly appropriate since the relations $x + ct = \kappa$ and $x - ct = \kappa$, where κ is a constant, are termed the characteristics of Eq. $\frac{\partial^2 w}{\partial t^2} = c^2 \frac{\partial^2 w}{\partial x^2}$. Note that at the points along the solution given by these characteristics, the solution has a constant value. For example, for the peak values in figure $\kappa = 0$ and the solution is constant at the peak value 0.5 for $x + ct = x - ct = 0$.

- The solution remains at zero for $x = x_l = 50$ so that the interval $-50 \le x \le 50$ is equivalent to the infinite interval $-\infty \le x \le \infty$.

The parameters that produced the numerical output in figure above are listed in table.

Table: Numerical values of parameters that produced the output of figure

Parameter	Description	Value
c	velocity in Eq. $\frac{\partial^2 w}{\partial t^2} = c^2 \frac{\partial^2 w}{\partial x^2}$	1
λ	constant in Eqs. $w(x, t=0) = e^{-\lambda x^2}$, $w_i^0 = e^{-\lambda x_i^2}$, $w(x,t) = \frac{1}{2}[e^{-\lambda(x+ct)^2} + e^{-\lambda(x-ct)^2}]$	0.05
N, xl	number of grid points and half length in *x*	201, 50
h, tf	grid spacing in *t*, final value of *t*	0.0025, 30

Additional parameters follow from the values in table. Thus, the grid spacing in x is $\Delta x = 2 \times 50/(201-1) = 0.5$. The number of steps in t taken along the solution is 30/0.0025=12000, which is large, but was selected to achieve good accuracy in the numerical solution.

To explore the accuracy of the numerical solution, we can consider the peak values of $w(x,t)$ in figure. This is a stringent test of the numerical solution since the curvature of the solution is greatest at these peaks. The results are summarized in table.

Table: Numerical values of w(x,t) at the peak values displayed in figure.

t	$-x, x$	Peak values
0	0	1
10	-10, 10	0.5006, 0.5006
20	-20, 20	0.5011, 0.5011
30	-30, 30	0.5015, 0.5015

Of course, the peak analytical values given by Eq. $w(x,t) = \frac{1}{2}[e^{-\lambda(x+ct)^2} + e^{-\lambda(x-ct)^2}]$ are $w(x=0, t=0) = 1$, $w(x = \pm t, t > 0) = 0.5$. Table indicates these peak values were attained within 0.5015 even for the largest value of t $(=30)$ so that errors did not accumulate excessively as the solution progressed through t. These errors could be reduced by increasing the number of grid points in x above $N = 201$. These errors could also presumably be reduced by using a more accurate (higher order) finite difference equation than the second order approximation of Eq. $\frac{d^2 w_i}{dx^2} \approx \frac{w_{i+1} - 2w_i + w_{i-1}}{\Delta x^2} + O(\Delta x^2)$.

This explicit finite difference numerical solution also has a stability limit (like the preceding parabolic problem). In this case

$$c\Delta t/\Delta x < 1.$$

Stability constraint in the above equation is the Courant–Friedrichs–Lewy (or CFL) condition. For the present numerical solution, $1 \times 0.0025/0.5 = 0.005$ so the CFL condition is easily satisfied. In other words, the parameters of Table were chosen primarily for accuracy and not stability.

Elliptic PDE

The elliptic PDE (Laplace's equation)

$$\frac{\partial^2 w}{\partial x^2} + \frac{\partial^2 w}{\partial y^2} = 0$$

has two boundary value independent variables, x and y, and no initial value variable. Thus, since the preceding numerical methods required an integration with respect to an initial value variable (t), and was accomplished by Euler's method, Eq. $w_i^{k+1} = w_i^k + \frac{dw_i^k}{dt} h + O(h^2)$, we cannot develop a numerical solution for Eq. $\frac{\partial^2 w}{\partial x^2} + \frac{\partial^2 w}{\partial y^2} = 0$ directly using these methods. Rather, we will convert Eq. $\frac{\partial^2 w}{\partial x^2} + \frac{\partial^2 w}{\partial y^2} = 0$

from an elliptic problem to an associated parabolic problem by adding a derivative in an initial value variable. Thus, the PDE we will now consider is

$$\frac{\partial w}{\partial t} = \frac{\partial^2 w}{\partial x^2} + \frac{\partial^2 w}{\partial y^2}$$

The idea then is to integrate Eq. $\frac{\partial w}{\partial t} = \frac{\partial^2 w}{\partial x^2} + \frac{\partial^2 w}{\partial y^2}$ forward in t until the numerical solution approaches the condition $\frac{\partial w}{\partial t} \to 0$ so that Eq. $\frac{\partial w}{\partial t} = \frac{\partial^2 w}{\partial x^2} + \frac{\partial^2 w}{\partial y^2}$ then reverts back to Eq. $\frac{\partial^2 w}{\partial x^2} + \frac{\partial^2 w}{\partial y^2} = 0$, i.e., the solution under this condition is for Eq. $\frac{\partial^2 w}{\partial x^2} + \frac{\partial^2 w}{\partial y^2} = 0$ as required.

Equation $\frac{\partial w}{\partial t} = \frac{\partial^2 w}{\partial x^2} + \frac{\partial^2 w}{\partial y^2}$ is first order in *t*, and second order in *x* and *y*. It therefore requires one IC and two BCs (for *x* and *y*). For the IC, since *t* has been added to the problem only to provide a solution to Eq. $\frac{\partial^2 w}{\partial x^2} + \frac{\partial^2 w}{\partial y^2} = 0$ through Eq. $\frac{\partial w}{\partial t} = \frac{\partial^2 w}{\partial x^2} + \frac{\partial^2 w}{\partial y^2}$ the choice of an IC is completely arbitrary (it is not part of the original problem). For the present analysis, we will use

$$w(x, y, t = 0) = \kappa$$

where κ is a constant to be selected (logically, it should be in the neighborhood of the expected solution to Eq. $\frac{\partial^2 w}{\partial x^2} + \frac{\partial^2 w}{\partial y^2} = 0$, but it's precise value is not critical to the success of the numerical method).

For the two BCs in *x*, we will use homogeneous (zero) Dirichlet BCs

$$w(x = 0, y, t) = 0$$

$$w(x = x_l, y, t) = 0$$

where x_l is the upper boundary value of x.

For the first BC in *y*, we will use a homogeneous Neumann BC

$$\frac{\partial w(x, y = 0, t)}{\partial y} = 0$$

For the second BC in *y*, we will use a nonhomogeneous Neumann BC

$$\frac{\partial w(x, y = y_l, t)}{\partial y} = \sin(\pi x)\pi \sinh(\pi y_l)$$

where y_l is the upper boundary value of y. Note that $\frac{\partial w(x,y=y_l)}{\partial y}$ in BC is a function of x.

The analytical solution to Eqs. $\frac{\partial^2 w}{\partial x^2}+\frac{\partial^2 w}{\partial y^2}=0$ (note, not Eq. $\frac{\partial w}{\partial t}=\frac{\partial^2 w}{\partial x^2}+\frac{\partial^2 w}{\partial y^2}$), $\frac{\partial w(x,y=y_l,t)}{\partial y}=\sin(\pi x)\pi\sinh(\pi y_l)$ is a special case of one of the solutions stated previously

$$w(x,y)=\sin(\pi x)\cos\text{h}(\pi y)$$

The analytical solution of Eq. $w(x,y)=\sin(\pi x)\cos\text{h}(\pi y)$ will be used to evaluate the numerical solution.

To develop a numerical solution to Eq. $\frac{\partial w}{\partial t}=\frac{\partial^2 w}{\partial x^2}+\frac{\partial^2 w}{\partial y^2}$, we first replace all of the derivatives with finite differences in analogy with the preceding numerical solutions. The positions in *x*, *y* and *t* will be denoted with indices*i*, *j* and *k*, respectively. The corresponding increments are Δx, Δy and h. Application of these ideas to Eq. $\frac{\partial w}{\partial t}=\frac{\partial^2 w}{\partial x^2}+\frac{\partial^2 w}{\partial y^2}$ gives the finite difference approximation

$$w_{i,j}^{k+1}=w_{i,j}^{k}+h\left(\frac{w_{i+lj}^{k}-2w_{i,j}^{k}+w_{i,\text{j}}^{k}}{\Delta x^2}+\frac{w_{i,j+1}^{k}-2w_{i,k}^{k}+w_{i,j-1}^{k}}{\Delta y^2}\right)$$

$$i=1,2,\ldots,Nx;\, j=1,2,\ldots,Ny;\quad k=1,2,\ldots$$

BCs ($w(x=0,y,t)=0$)–($w(x=x_l,y,t)=0$) in the difference notation are

$$w_{l,\text{j}}^{k}=0$$

$$w_{Nx,\text{j}}^{k}=0$$

BC $\frac{\partial w(x,y=0,t)}{\partial y}=0$ in difference notation is

$$w_{i,0}^{k}=w_{i,2}^{k}$$

where $w_{i,0}^{k}$ is a fictitious value that can be used in Eq.

$i=1,2,\ldots,Nx;\, j=1,2,\ldots,Ny;\quad k=1,2,\ldots$ for *j*=1.

BC $\frac{\partial w(x,y=y_l,t)}{\partial y}=\sin(\pi x)\pi\sinh(\pi y_l)$ in difference notation is

$$w_{i,N_y+1}^{k} = w_{i,N_y-1}^{k} + 2\Delta y \sin(\pi x_i)\pi \sinh(\pi y_{N_y})$$

where $w_{i,N_y+1}^{k}\Big|$ is a fictitious value that can be used in Eq.

$i = 1, 2, \ldots, Nx;\ j = 1, 2, \ldots, Ny;\quad k = 1, 2, \ldots$ for j=N_y.

References

- Differential-equation, science: britannica.com, Retrieved 16 February, 2019
- OrdinaryDifferentialEquation: wolfram.com, Retrieved 5 April, 2019
- Ordinary_differential_equation: cs.mcgill.ca, Retrieved 26 January, 2019
- Partial_differential_equation: scholarpedia.org, Retrieved 1 August, 2019

7

Probability and Statistics

Probability is a measure that quantifies the likelihood of events that might occur. Statistics is a subset of mathematics which deals with data collection, analysis, organization, interpretation and presentation. The chapter closely examines the key concepts of probability and statistics to provide an extensive understanding of mathematics.

Probability

Probability is the likelihood of something happening. When someone tells you the probability of something happening, they are telling you how likely that something is. When people buy lottery tickets, the probability of winning is usually stated, and sometimes, it can be something like 1/10,000,000 (or even worse). This tells you that it is not very likely that you will win.

The formula for probability tells you how many choices you have over the number of possible combinations.

Calculating Probability

To calculate probability, you need to know how many possible options or outcomes there are and how many right combinations you have. Let's calculate the probability of throwing dice, and how it works.

First, we know that a die has a total of 6 possible outcomes. You can roll a 1, 2, 3, 4, 5, or 6. Next, we need to know how many choices we have. Whenever you roll, you will get one of the numbers. You can't roll and get two different numbers with one die. So, our number of choices is 1. Using our formula for probability, we get a probability of 1/6.

$$probability = \frac{possible\ choices}{total\ number\ of\ options}$$

Our probability of rolling any of the numbers is 1/6. The probability of rolling a 2 is 1/6, of rolling a 3 is also 1/6, and so on.

Let's try another problem. Let's say we have a grab bag of apples and oranges. We want to find out the probability of picking an apple from the bag. One thing we need to know is the number of apples in the bag because that gives us the number of 'correct' choices, which is the number of our possible choices in the top part of the calculation.

$$probability = \frac{1}{6}$$

We also need to know the total number of fruits in the bag, for this gives us the total number of choices we have, or the total number of options in the bottom part of the calculation. The person with the grab bag tells us there are 10 apples and 20 oranges in the bag. So, what is our probability of picking an apple? We have 10 apples, one of which we want, and a total of 30 fruits to pick from.

$$probability = \frac{10}{30}$$
$$= \frac{1}{3}$$

Our probability is 1/3 for picking an apple. If you compare this with our probability of rolling a number on a die, the probability of picking an apple from the grab bag is higher. It is more likely that we will pick an apple than that we will roll a particular number.

In both cases, we can leave the probability in fraction form or we can convert it to decimal form: 1/6 becomes 0.17, and 1/3 becomes 0.33.

Conditional Probability

Conditional probability is the probability of one event occurring with some relationship to one or more other events. For example:

- Event A is that it is raining outside, and it has a 0.3 (30%) chance of raining today.
- Event B is that you will need to go outside, and that has a probability of 0.5 (50%).

A conditional probability would look at these two events in relationship with one another, such as the probability that it is both raining and you will need to go outside.

The formula for conditional probability is,

$$P(B|A) = P(A \text{ and } B) / P(A)$$

which you can also rewrite as,

$$P(B|A) = P(A \cap B) / P(A).$$

Conditional Probability Formula Examples

Example

In a group of 100 sports car buyers, 40 bought alarm systems, 30 purchased bucket seats, and 20 purchased an alarm system and bucket seats. If a car buyer chosen at random bought an alarm system, what is the probability they also bought bucket seats.

Step 1: Figure out P(A). It's given in the question as 40%, or 0.4.

Step 2: Figure out P(A∩B). This is the intersection of A and B: both happening together. It's given in the question 20 out of 100 buyers, or 0.2.

Step 3: Insert your answers into the formula:

$$P(B|A)=P(A\cap B)/P(A)=0.2/0.4=0.5.$$

The probability that a buyer bought bucket seats, given that they purchased an alarm system, is 50%.

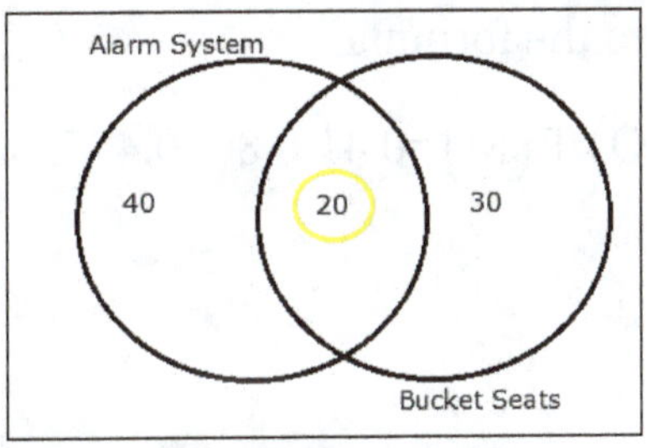

Venn diagram showing that 20 out of 40 alarm buyers purchased bucket seats.

Example

This question uses the following contingency table:

	Have pets	Dot not have pets	Total
Male	0.41	0.08	0.49
Female	0.45	0.06	0.51
Total	0.86	0.14	1

What is the probability a randomly selected person is male, given that they own a pet?

Step 1: Repopulate the formula with new variables so that it makes sense for the question (optional, but it helps to clarify what you're looking for). I'm going to say M is for male and PO stands for pet owner, so the formula becomes,

$$P(M|PO)=P(M\cap PO)/P(PO)$$

Step 2: Figure out P(M∩PO) from the table. The intersection of male/pets (the intersection on the table of these two factors) is 0.41.

	Have pets	Dot not have pets	Total
Male	0.41	0.08	0.49
Female	0.45	0.06	0.51
Total	0.86	0.14	1

Step 3: Figure out P(PO) from the table. From the total column, 86% (0.86) of respondents had a pet.

	Have pets	Dot not have pets	Total
Male	0.41	0.08	0.49
Female	0.45	0.06	0.51
Total	0.86	0.14	1

Step 4: Insert your values into the formula:

$$P(M|PO)=P(M\cap PO)/P(M)=0.41/0.86=0.477, \text{ or } 47.7\%.$$

Single Events

Example

There are 6 beads in a bag, 3 are red, 2 are yellow and 1 is blue. What is the probability of picking a yellow?

The probability is the number of yellows in the bag divided by the total number of balls, i.e. 2/6 = 1/3.

Example

There is a bag full of coloured balls, red, blue, green and orange. Balls are picked out and replaced. John did this 1000 times and obtained the following results-

- Number of blue balls picked out: 300
- Number of red balls: 200
- Number of green balls: 450
- Number of orange balls: 50

a) What is the probability of picking a green ball?

b) For every 1000 balls picked out, 450 are green. Therefore P(green) = 450/1000 = 0.45.

If there are 100 balls in the bag, how many of them are likely to be green?

The experiment suggests that 450 out of 1000 balls are green. Therefore, out of 100 balls, 45 are green (using ratios).

Multiple Events

Independent and Dependent Events

Suppose now we consider the probability of 2 events happening. For example, we might throw 2 dice and consider the probability that both are 6's.

We call two events independent if the outcome of one of the events doesn't affect the outcome of another. For example, if we throw two dice, the probability of getting a 6 on the second die is the same, no matter what we get with the first one- it's still 1/6.

On the other hand, suppose we have a bag containing 2 red and 2 blue balls. If we pick 2 balls out of the bag, the probability that the second is blue depends upon what the colour of the first ball picked was. If the first ball was blue, there will be 1 blue and 2 red balls in the bag when we pick the second ball. So the probability of getting a blue is 1/3. However, if the first ball was red, there will be 1 red and 2 blue balls left so the probability the second ball is blue is 2/3. When the probability of one event depends on another, the events are dependent.

Possibility Spaces

When working out what the probability of two things happening is, a probability/ possibility space can be drawn. For example, if you throw two dice, what is the probability that you will get: a) 8, b) 9, c) either 8 or 9?

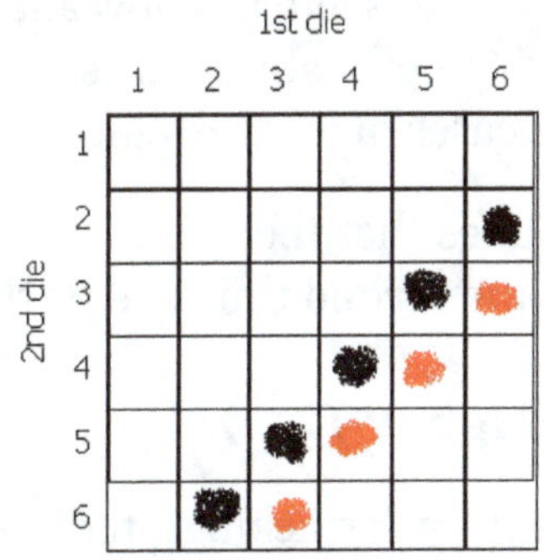

a) The black blobs indicate the ways of getting 8 (a 2 and a 6, a 3 and a 5, ...). There are 5 different ways. The probability space shows us that when throwing 2 dice, there are 36 different possibilities (36 squares). With 5 of these possibilities, you will get 8. Therefore P(8) = 5/36 .

b) The red blobs indicate the ways of getting 9. There are four ways, therefore P(9) = 4/36 = 1/9.

c) You will get an 8 or 9 in any of the 'blobbed' squares. There are 9 altogether, so P(8 or 9) = 9/36 = 1/4 .

Probability Trees

Another way of representing 2 or more events is on a probability tree.

Example

There are 3 balls in a bag: red, yellow and blue. One ball is picked out, and not replaced, and then another ball is picked out.

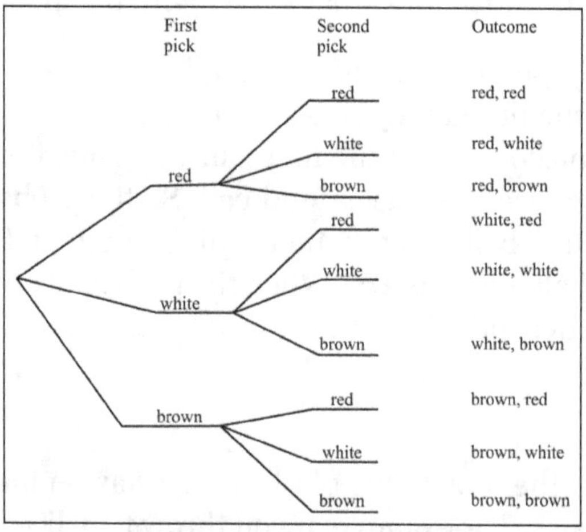

The first ball can be red, yellow or blue. The probability is 1/3 for each of these. If a red ball is picked out, there will be two balls left, a yellow and blue. The probability the second ball will be yellow is 1/2 and the probability the second ball will be blue is 1/2. The same logic can be applied to the cases of when a yellow or blue ball is picked out first.

In this example, the question states that the ball is not replaced. If it was, the probability of picking a red ball (etc.) the second time will be the same as the first (i.e. 1/3).

The AND and OR rules (Higher Tier)

In the above example, the probability of spicking a red first is 1/3 and a yellow second is 1/2. The probability that a red AND then a yellow will be picked is 1/3 × 1/2 = 1/6 (this is shown at the end of the branch). The rule is:

- If two events A and B are independent (this means that one event does not depend on the other), then the probability of both A and B occurring is found by multiplying the probability of A occurring by the probability of B occurring.

The probability of picking a red OR yellow first is 1/3 + 1/3 = 2/3. The rule is:

- If we have two events A and B and it isn't possible for both events to occur, then the probability of A or B occuring is the probability of A occurring + the probability of B occurring.

On a probability tree, when moving from left to right we multiply and when moving down we add.

Example

What is the probability of getting a yellow and a red in any order?

This is the same as: what is the probability of getting a yellow AND a red OR a red AND a yellow.

P(yellow and red) = 1/3 × 1/2 = 1/6

P(red and yellow) = 1/3 × 1/2 = 1/6

P(yellow and red or red and yellow) = 1/6 + 1/6 = 1/3

Probability Theory

Probability theory is a branch of mathematics concerned with the analysis of random phenomena. The outcome of a random event cannot be determined before it occurs, but it may be any one of several possible outcomes. The actual outcome is considered to be determined by chance.

The word probability has several meanings in ordinary conversation. Two of these are particularly important for the development and applications of the mathematical theory of probability. One is the interpretation of probabilities as relative frequencies, for which simple games involving coins, cards, dice, and roulette wheels provide examples. The distinctive feature of games of chance is that the outcome of a given trial cannot be predicted with certainty, although the collective results of a large number of trials display some regularity. For example, the statement that the probability of "heads" in tossing a coin equals one-half, according to the relative frequency interpretation, implies that in a large number of tosses the relative frequency with which "heads" actually occurs will be approximately one-half, although it contains no implication concerning the outcome of any given toss. There are many similar examples involving groups of people, molecules of a gas, genes, and so on. Actuarial statements about the life expectancy for persons of a certain age describe the collective experience of a large number of individuals but do not purport to say what will happen to any particular person. Similarly, predictions about the chance of a genetic disease occurring in a child of parents having a known genetic makeup are statements about relative frequencies of occurrence in a large number of cases but are not predictions about a given individual.

Random Variables, Distributions, Expectation and Variance

Random Variables

Usually it is more convenient to associate numerical values with the outcomes of an experiment than to work directly with a nonnumerical description such as "red ball on the first draw." For example, an outcome of the experiment of drawing n balls with replacement from an urn containing black and red balls is an n-tuple that tells us whether a red or a black ball was drawn on each of the draws. This n-tuple is conveniently represented by an n-tuple of ones and zeros, where the appearance of a one in the kth position indicates that a red ball was drawn on the kth draw. A quantity of particular interest is the number of red balls drawn, which is just the sum of the entries in this numerical description of the experimental outcome. Mathematically a rule that associates with every element of a given set a unique real number is called a "(real-valued) function." In the history of statistics and probability, real-valued functions defined on a sample space have traditionally been called "random variables." Thus, if a sample space S has the generic element e, the outcome of an experiment, then a random variable is a real-valued function X = X(e). Customarily one omits the argument e in the notation for a random variable. For the experiment of drawing balls from an urn containing black and red balls, R, the number of red balls drawn, is a random variable. A particularly useful random variable is 1[A], the indicator variable of the event A, which equals 1 if A occurs and 0 otherwise. A "constant" is a trivial random variable that always takes the same value regardless of the outcome of the experiment.

Probability Distribution

Suppose X is a random variable that can assume one of the values $x_1, x_2, \ldots, x_m$, according to the outcome of a random experiment, and consider the event $\{X = x_i\}$, which is a shorthand notation for the set of all experimental outcomes e such that $X(e) = x_i$. The probability of this event, $P\{X = x_i\}$, is itself a function of x_i, called the probability distribution function of X. Thus, the distribution of the random variable R defined is the function of i = 0, 1,..., n given in the binomial equation. Introducing the notation $f(x_i) = P\{X = x_i\}$, one sees from the basic properties of probabilities that,

$$f(x_i) \geq 0 \text{ for all } i \sum_i f(x_i) = 1,$$

and

$$P\{a < X \leq b\} = \sum_{a < x_i \leq b} f(x_i)$$

for any real numbers a and b. If Y is a second random variable defined on the same sample space as X and taking the values $y_1, y_2, \ldots, y_n$, the function of two variables $h(x_i, y_j) = P\{X = x_i, Y = y_j\}$ is called the joint distribution of X and Y. Since $\{X = x_i\} = \cup_j\{X = x_i, Y = y_j\}$, and this union consists of disjoint events in the sample space,

$$f(x_i) = \sum_j h(x_i, y_j), \quad \text{for all } i.$$

Often f is called the marginal distribution of X to emphasize its relation to the joint distribution of X and Y. Similarly, $g(y_j) = \Sigma_i h(x_i, y_j)$ is the (marginal) distribution of Y. The random variables X and Y are defined to be independent if the events $\{X = x_i\}$ and $\{Y = y_j\}$ are independent for all i and j—i.e., if $h(x_i, y_j) = f(x_i)g(y_j)$ for all i and j. The joint distribution of an arbitrary number of random variables is defined similarly.

Suppose two dice are thrown. Let X denote the sum of the numbers appearing on the two dice, and let Y denote the number of even numbers appearing. The possible values of X are 2, 3,..., 12, while the possible values of Y are 0, 1, 2. Since there are 36 possible outcomes for the two dice, the accompanying table giving the joint distribution h(i, j) (i = 2, 3,..., 12; j = 0, 1, 2) and the marginal distributions f(i) and g(j) is easily computed by direct enumeration.

For more complex experiments, determination of a complete probability distribution usually requires a combination of theoretical analysis and empirical experimentation and is often very difficult. Consequently, it is desirable to describe a distribution insofar as possible by a small number of parameters that are comparatively easy to evaluate and interpret. The most important are the mean and the variance. These are both defined in terms of the "expected value" of a random variable.

Expected Value

Given a random variable X with distribution f, the expected value of X, denoted E(X), is defined by $E(X) = \Sigma_i x_i f(x_i)$. In words, the expected value of X is the sum of each of the possible values of X multiplied by the probability of obtaining that value. The expected value of X is also called the mean of the distribution f. The basic property of E is that of linearity: if X and Y are random variables and if a and b are constants, then E(aX+bY) = aE(X)+bE(Y). To see why this is true, note that aX + bY is itself a random variable, which assumes the values ax_i+by_j with the probabilities $h(x_i, y_j)$. Hence,

$$E(aX+bY) = \sum_{i,j}(ax_i + by_j)h(x_i, y_j)$$
$$= a\sum_{i,j} x_i h(x_i, y_j) + b\sum_{i,j} y_j h(x_i, y_j)$$

If the first sum on the right-hand side is summed over j while holding i fixed, by equation the result is,

$$f(x_i) = \sum_j h(x_i, y_j), \text{for all } i.$$

$$\sum_i x_i f(x_i),$$

which by definition is E(X). Similarly, the second sum equals E(Y).

If 1[A] denotes the "indicator variable" of A—i.e., a random variable equal to 1 if A occurs and equal to 0 otherwise—then E{1[A]}=1×P(A)+0×P(A^c)=P(A). This shows that the concept of expectation includes that of probability as a special case.

Consider the number R of red balls in n draws with replacement from an urn containing a proportion p of red balls. From the definition and the binomial distribution of R,

$$E(R) = \sum_i i\binom{n}{i} p^i q^{n-i},$$

Which can be evaluated by algebraic manipulation and found to equal np. It is easier to use the representation R = 1[A_1] +⋯+ 1[A_n], where A_k denotes the event "the kth draw results in a red ball." Since E{1[A_k]} = p for all k, by linearity E(R) = E{1[A_1]} +⋯+ E{1[A_n]} = np. This argument illustrates the principle that one can often compute the expected value of a random variable without first computing its distribution. For another example, suppose n balls are dropped at random into n boxes. The number of empty boxes, Y, has the representation Y = 1[B_1] +⋯+ 1[B_n], where Bk is the event that "the kth box is empty." Since the kth box is empty if and only if each of the n balls went into one of the other n – 1 boxes, P(B_k) = [(n – 1)/n]n for all k, and consequently E(Y) = n(1 – 1/n)$_n$. The exact distribution of Y is very complicated, especially if n is large.

Many probability distributions have small values of f(x_i) associated with extreme (large or small) values of x_i and larger values of f(x_i) for intermediate x_i. For example, both marginal distributions in the table are symmetrical about a midpoint that has relatively high probability, and the probability of other values decreases as one moves away from the midpoint. Insofar as a distribution f(x_i) follows this kind of pattern, one can interpret the mean of f as a rough measure of location of the bulk of the probability distribution, because in the defining sum the values xi associated with large values of f(x_i) more or less define the centre of the distribution. In the extreme case, the expected value of a constant random variable is just that constant.

Variance

It is also of interest to know how closely packed about its mean value a distribution is. The most important measure of concentration is the variance, denoted by Var(X) and defined by Var(X) = E{[X – E(X)]^2}. By linearity of expectations, one has equivalently Var(X) = E(X^2) – {E(X)}^{2}. The standard deviation of X is the square root of its variance. It has a more direct interpretation than the variance because it is in the same units as X. The variance of a constant random variable is 0. Also, if c is a constant, Var(cX) = c^2Var(X).

There is no general formula for the expectation of a product of random variables. If the random variables X and Y are independent, E(XY) = E(X)E(Y). This can be used to show that, if $X_1,\ldots,X_n$ are independent random variables, the variance of the sum $X_1+\cdots+X_n$ is just the sum of the individual variances, $Var(X_1)+\cdots+Var(X_n)$. If the Xs have the same distribution and are independent, the variance of the average $(X_1+\cdots+X_n)/n$ is $Var(X_1)/n$. Equivalently, the standard deviation of $(X_1+\cdots+X_n)/n$ is the standard deviation of X_1 divided by Square root of $\sqrt{n}$. This quantifies the intuitive notion that the average of repeated observations is less variable than the individual observations. More precisely, it says that the variability of the average is inversely proportional to the square root of the number of observations. This result is tremendously important in problems of statistical inference.

Consider again the binomial distribution equation. As in the calculation of the mean value, one can use the definition combined with some algebraic manipulation to show that, if R has the binomial distribution, then Var(R) = npq. From the representation $R = 1[A_1]+\cdots+1[A_n]$ defined above, and the observation that the events A_k are independent and have the same probability, it follows that,

$$\text{Var}(R) = \text{Var}\{1[A_1]\} + \ldots +_n \text{Var}\{1[A_n]\} = \text{Var}\{1[A_1]\}.$$

Moreover,

$$\text{Var}\{1[A_1]\} = E\{1[A_1]^2\} - [E\{1[A_n]\}]^2 = p - p^2 = pq.$$

so Var(R) = npq.

The conditional distribution of Y given $X = x_i$ is defined by:

$$P\{Y = y_j | X = x_i\} = \frac{h(x_i, y_j)}{f(x_i)}$$

(compare Bayes's theorem), and the conditional expectation of Y given $X = x_i$ is

$$E(Y|X = x_i) = \sum_j \frac{y_j h(x_i, y_j)}{f(x_i)}$$

One can regard E(Y|X) as a function of X; since X is a random variable, this function of X must itself be a random variable. The conditional expectation E(Y|X) considered as a random variable has its own (unconditional) expectation E{E(Y|X)}, which is calculated by multiplying equation $E(Y|X = x_i) = \sum_j \frac{y_j h(x_i, y_j)}{f(x_i)}$ by $f(x_i)$ and summing over i to obtain the important formula

$$E\{E(Y|X)\} = E(Y).$$

Properly interpreted, equation $E\{E(Y|X)\} = E(Y)$. is a generalization of the law of total probability.

For a simple example of the use of equation $E\{E(Y|X)\} = E(Y)$, recall the problem of the gambler's ruin and let e(x) denote the expected duration of the game if Peter's fortune is initially equal to x. The reasoning leading to equation in conjunction with equation $E\{E(Y|X)\} = E(Y)$ shows that e(x) satisfies the equations $e(x)=1+pe(x+1)+qe(x-1)$ for $x=1,2,\ldots,m-1$ with the boundary conditions e(0) = e(m) = 0. The solution for p ≠ 1/2 is rather complicated; for p = 1/2, e(x) = x(m − x).

An Alternative Interpretation of Probability

In ordinary conversation the word probability is applied not only to variable phenomena but also to propositions of uncertain veracity. The truth of any proposition concerning the outcome of an experiment is uncertain before the experiment is performed. Many other uncertain propositions cannot be defined in terms of repeatable experiments. An individual can be uncertain about the truth of a scientific theory, a religious doctrine, or even about the occurrence of a specific historical event when inadequate or conflicting eyewitness accounts are involved. Using probability as a measure of uncertainty enlarges its domain of application to phenomena that do not meet the requirement of repeatability. The concomitant disadvantage is that probability as a measure of uncertainty is subjective and varies from one person to another.

According to one interpretation, to say that someone has subjective probability p that a proposition is true means that for any integers r and b with r/(r + b) < p, if that individual is offered an opportunity to bet the same amount on the truth of the proposition or on "red in a single draw" from an urn containing r red and b black balls, he prefers the first bet, while, if r/(r + b) > p, he prefers the second bet.

An important stimulus to modern thought about subjective probability has been an attempt to understand decision making in the face of incomplete knowledge. It is assumed that an individual, when faced with the necessity of making a decision that may have different consequences depending on situations about which he has incomplete knowledge, can express his personal preferences and uncertainties in a way consistent with certain axioms of rational behaviour. It can then be deduced that the individual has a utility function, which measures the value to him of each course of action when each of the uncertain possibilities is the true one, and a "subjective probability distribution," which expresses quantitatively his beliefs about the uncertain situations. The individual's optimal decision is the one that maximizes his expected utility with respect to his subjective probability. The concept of utility goes back at least to Daniel Bernoulli (Jakob Bernoulli's nephew) and was developed in the 20th century by John von Neumann and Oskar Morgenstern, Frank P. Ramsey, and Leonard J. Savage, among others. Ramsey and Savage stressed the importance of subjective probability as a concomitant ingredient of decision making in the face of uncertainty. An alternative approach to subjective probability without the use of utility theory was developed by Bruno de Finetti.

The mathematical theory of probability is the same regardless of one's interpretation of the concept, although the importance attached to various results can depend very much on the interpretation. In particular, in the theory and applications of subjective probability, Bayes's theorem plays an important role.

For example, suppose that an urn contains N balls, r of which are red and b = N − r of which are black, but r (hence b) is unknown. One is permitted to learn about the value of r by performing the experiment of drawing with replacement n balls from the urn. Suppose also that one has a subjective probability distribution giving the probability f(r) that the number of red balls is in fact r where f(0) +⋯+ f(N) = 1. This distribution is called an a priori distribution because it is specified prior to the experiment of drawing balls from the urn. The binomial distribution is now a conditional probability, given the value of r. Finally, one can use Bayes's theorem to find the conditional probability that the unknown number of red balls in the urn is r, given that the number of red balls drawn from the urn is i. The result is,

$$\frac{f(r)r^i b^{n-1}}{\sum_{r_0=0}^{n} f(r_0)\,\mathrm{r}_0^i\, b_0^{n-1}}, \quad \text{where } \mathrm{b}_0 = N - r_0.$$

This distribution, derived by using Bayes's theorem to combine the a priori distribution with the conditional distribution for the outcome of the experiment, is called the a posteriori distribution.

The virtue of this calculation is that it makes possible a probability statement about the composition of the urn, which is not directly observable, in terms of observable data, from the composition of the sample taken from the urn. The weakness, as indicated above, is that different people may choose different subjective probabilities for the composition of the urn a priori and hence reach different conclusions about its composition a posteriori.

To see how this idea might apply in practice, consider a simple urn model of opinion polling to predict which of two candidates will win an election. The red balls in the urn are identified with voters who will vote for candidate A and the black balls with those voting for candidate B. Choosing a sample from the electorate and asking their preferences is a well-defined random experiment, which in theory and in practice is repeatable. The composition of the urn is uncertain and is not the result of a well-defined random experiment. Nevertheless, to the extent that a vote for a candidate is a vote for a political party, other elections provide information about the content of the urn, which, if used judiciously, should be helpful in supplementing the results of the actual sample to make a prediction. Exactly how to use this information is a difficult problem in which individual judgment plays an important part. One possibility is to incorporate the prior information into an a priori distribution about the electorate, which is then combined via Bayes's theorem with the outcome of the sample and summarized by an a posteriori distribution.

Bayes' Theorem

Bayes' theorem is a mathematical equation used in probability and statistics to calculate conditional probability. In other words, it is used to calculate the probability of an event based on its association with another event. The theorem is also known as Bayes' law or Bayes' rule.

Formula for Bayes' Theorem

There are several different ways to write the formula for Bayes' theorem. The most common form is:

$$P(A|B) = P(B|A)P(A) / P(B)$$

where A and B are two events and $P(B) \neq 0$

- $P(A \mid B)$ is the conditional probability of event A occurring given that B is true.
- $P(B \mid A)$ is the conditional probability of event B occurring given that A is true.

P(A) and P(B) are the probabilities of A and B occurring independently of one another (the marginal probability).

Proof of Bayes' Theorem

The probability of two events A and B happening, $P(A \cap B)$, is the probability of A, P(A), times the probability of B given that A has occurred, $P(B|A)$.

$$P(A \cap B) = P(A)P(B|A)$$

The probability of two events A and B happening, $P(A \cap B)$, is the probability of A, P(A), times the probability of B given that A has occurred, $P(B|A)$.

$$P(A \cap B) = P(B)P(A|B)$$

Equating the two yields,

$$P(B)P(A|B) = P(A)P(B|A)$$

and thus,

$$P(A|B) = P(A)\frac{P(B|A)}{P(B)}$$

This equation, known as Bayes Theorem is the basis of statistical inference.

Proof of Bayes Theorem

The probability of two events A and B happening, P(A∩B), is the probability of A, P(A), times the probability of B given that A has occurred, P(B|A).

$$P(A \cap B) = P(A)P(B|A)$$

On the other hand, the probability of A and B is also equal to the probability of B times the probability of A given B.

$$P(A \cap B) = P(B)P(A|B)$$

Equating the two yields:

$$P(B)P(A|B) = P(A)P(B|A)$$

and thus

$$P(A|B) = P(A)\frac{P(B|A)}{P(B)}$$

This equation, known as Bayes Theorem is the basis of statistical inference.

Statistics

Statistics is the study of the collection, analysis, interpretation, presentation, and organization of data. In other words, it is a mathematical discipline to collect, summarize data.

Mathematical Statistics

Mathematical statistics is the application of mathematics to statistics, which was originally conceived as the science of the state — the collection and analysis of facts about a country: its economy, and, military, population, and so forth.

Mathematical techniques used for this include mathematical analysis, linear algebra, stochastic analysis,differential equation and measure-theoretic probability theory.

Scope

Statistics is used in many sectors such as psychology, geology, sociology, weather forecasting, probability and much more. The goal of statistics is to gain understanding from data it focuses on applications and hence, it is distinctively considered as a Mathematical science.

Methods

The methods of collecting, summarizing, analyzing, and interpreting variable numerical data. Here are some of the methods provided below.

- Data collection
- Data summarization
- Statistical analysis

Data

Data is a collection of facts, such as numbers, words, measurements, observations etc.

Types of Data

- Qualitative data- it is descriptive data.

  Example- She can run fast, He is thin.

- Quantitative data- it is numerical information.

  Example- An Octopus is an Eight legged creature.

Types of Quantitative Data

- Discrete data- has a particular fixed value.It can be counted.
- Continuous data- is not fixed but has a range of data.It can be measured.

Representation of Data

Bar Graph

A Bar Graph represent grouped data with rectangular bars with lengths proportional to the values that they represent. The bars can be plotted vertically or horizontally.

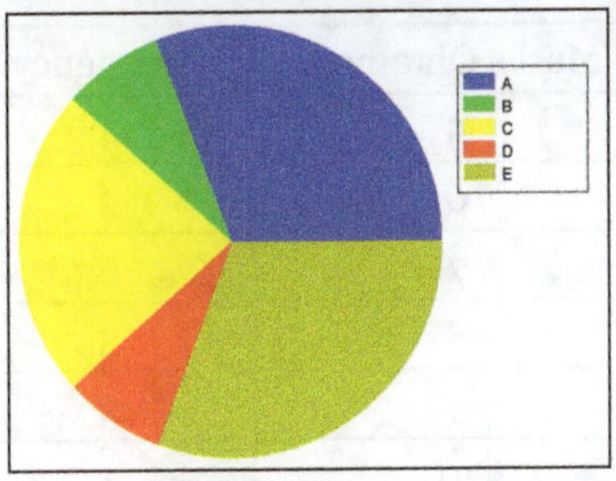

Pie Chart

A type of graph in which a circle is divided into Sectors that each represent a proportion of the whole.

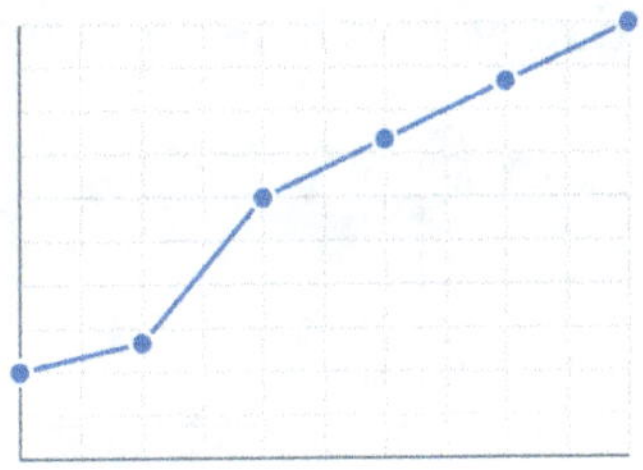

Line graph

The line chart is represented by a series of data-points connected with a straight line.

The series of data points are called 'markers.'

A pictorial symbol for a word or phrase, i.e. showing data with the help of pictures.Such as Apple, Banana & Cherry can have different number, it is just a representation of data.

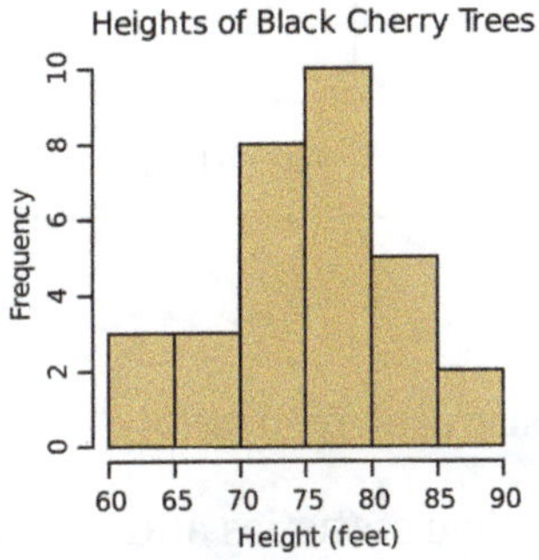

Histogram

A diagram consisting of rectangles whose area is proportional to the frequency of a variable and whose width is equal to the class interval.

Marks Obtained	Frequency
5	4
6	3
7	6
8	5
9	3
10	1

Frequency Distribution

The frequency of a data value is often represented by "f." A frequency table is constructed by arranging collected data values in ascending order of magnitude with their corresponding frequencies.

Sample Mean ($\bar{x}$)	$\frac{\Sigma x}{n}$
Population Mean (μ)	$\frac{\Sigma x}{N}$
Sample Standard Deviation (s)	$\sqrt{\frac{\Sigma(x-\bar{x})^2}{n-1}}$
Population Standard Deviation (σ)	$\sigma = \sqrt{\frac{(x-\mu)^2}{N}}$
Sample Variance (s^2)	$s^2 = \frac{\Sigma(x_i - \bar{x})}{n-1}$
Population Variance (σ^2)	$\sigma^2 = \frac{\Sigma(x_i - \bar{x})}{N} <$
Range (R)	Largest data value – smallest data value

Application

Some of the application of statistic are given below:

- Applied statistics, theoretical statistics and mathematical statistics

- Machine learning and data mining
- Statistics in society
- Statistical computing
- Statistics applied to mathematics or the arts.

Descriptive Statistics

Descriptive statistics are used to describe the basic features of the data in a study. They provide simple summaries about the sample and the measures. Together with simple graphics analysis, they form the basis of virtually every quantitative analysis of data.

Descriptive statistics are typically distinguished from inferential statistics. With descriptive statistics you are simply describing what is or what the data shows. With inferential statistics, you are trying to reach conclusions that extend beyond the immediate data alone. For instance, we use inferential statistics to try to infer from the sample data what the population might think. Or, we use inferential statistics to make judgments of the probability that an observed difference between groups is a dependable one or one that might have happened by chance in this study. Thus, we use inferential statistics to make inferences from our data to more general conditions; we use descriptive statistics simply to describe what's going on in our data.

Descriptive Statistics are used to present quantitative descriptions in a manageable form. In a research study we may have lots of measures. Or we may measure a large number of people on any measure. Descriptive statistics help us to simplify large amounts of data in a sensible way. Each descriptive statistic reduces lots of data into a simpler summary.

Measures of Central Tendency

A measure of central tendency is a single value that attempts to describe a set of data by identifying the central position within that set of data. As such, measures of central tendency are sometimes called measures of central location. They are also classed as summary statistics. The mean (often called the average) is most likely the measure of central tendency that you are most familiar with, but there are others, such as the median and the mode.

The mean, median and mode are all valid measures of central tendency, but under different conditions, some measures of central tendency become more appropriate to use than others.

Mean (Arithmetic)

The mean (or average) is the most popular and well known measure of central tendency. It can be used with both discrete and continuous data, although its use is most often with continuous data. The mean is equal to the sum of all the values in the data set divided by the number of values in the data set. So, if we have n values in a data set and they have values $x_1, x_2, ..., x_n$, the sample mean, usually denoted by $\bar{x}$, is:

$$\bar{x} = \frac{(x_1 + x_2 + ... + x_n)}{n}$$

This formula is usually written in a slightly different manner using the Greek capitol letter, $\sum$., pronounced "sigma", which means "sum of":

$$\bar{x} = \frac{\sum x}{n}$$

You may have noticed that the above formula refers to the sample mean. It is called Simple Mean beacause because, in statistics, samples and populations have very different meanings and these differences are very important, even if, in the case of the mean, they are calculated in the same way. To acknowledge that we are calculating the population mean and not the sample mean, we use the Greek lower case letter "mu", denoted as μ:

$$\mu = \frac{\sum x}{n}$$

The mean is essentially a model of your data set. It is the value that is most common. You will notice, however, that the mean is not often one of the actual values that you have observed in your data set. However, one of its important properties is that it minimises error in the prediction of any one value in your data set. That is, it is the value that produces the lowest amount of error from all other values in the data set.

An important property of the mean is that it includes every value in your data set as part of the calculation. In addition, the mean is the only measure of central tendency where the sum of the deviations of each value from the mean is always zero.

Restrictions to using the Mean

The mean has one main disadvantage: it is particularly susceptible to the influence of outliers. These are values that are unusual compared to the rest of the data set by being especially small or large in numerical value. For example, consider the wages of staff at a factory below:

Staff	1	2	3	4	5	6	7	8	9	10
Salary	15*k*	18*k*	16*k*	14*k*	15*k*	15*k*	12*k*	17*k*	90*k*	95*k*

The mean salary for these ten staff is $30.7k. However, inspecting the raw data suggests that this mean value might not be the best way to accurately reflect the typical salary of a worker, as most workers have salaries in the $12k to 18k range. The mean is being skewed by the two large salaries. Therefore, in this situation, we would like to have a better measure of central tendency in this situation.

Another time when we usually prefer the median over the mean (or mode) is when our data is skewed (i.e., the frequency distribution for our data is skewed). If we consider the normal distribution - as this is the most frequently assessed in statistics - when the data is perfectly normal, the mean, median and mode are identical. Moreover, they all represent the most typical value in the data set. However, as the data becomes skewed the mean loses its ability to provide the best central location for the data because the skewed data is dragging it away from the typical value. However, the median best retains this position and is not as strongly influenced by the skewed values.

Median

The median is the middle score for a set of data that has been arranged in order of magnitude. The median is less affected by outliers and skewed data. In order to calculate the median, suppose we have the data below:

65	55	89	56	35	14	56	55	87	45	92

We first need to rearrange that data into order of magnitude (smallest first):

14	35	45	55	55	56	56	65	87	89	92

Our median mark is the middle mark - in this case, 56 (highlighted in bold). It is the middle mark because there are 5 scores before it and 5 scores after it. This works fine when you have an odd number of scores, but what happens when you have an even number of scores? What if you had only 10 scores? Well, you simply have to take the middle two scores and average the result. So, if we look at the example below:

65	55	89	56	35	14	56	55	87	45

We again rearrange that data into order of magnitude (smallest first):

14	35	45	55	55	56	56	65	87	89

Only now we have to take the 5th and 6th score in our data set and average them to get a median of 55.5.

Mode

The mode is the most frequent score in our data set. On a histogram it represents the highest bar in a bar chart or histogram. You can, therefore, sometimes consider the mode as being the most popular option. An example of a mode is presented below:

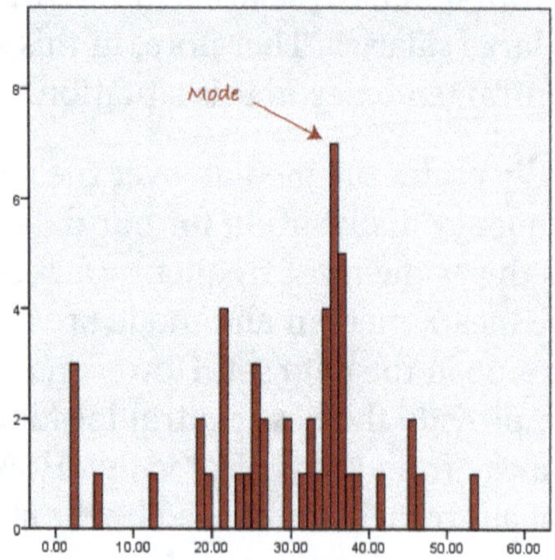

Normally, the mode is used for categorical data where we wish to know which is the most common category, as illustrated below:

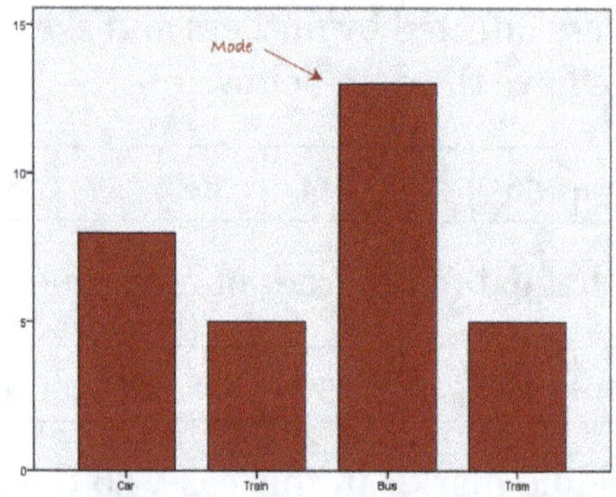

We can see above that the most common form of transport, in this particular data set, is the bus. However, one of the problems with the mode is that it is not unique, so it leaves us with problems when we have two or more values that share the highest frequency, such as below:

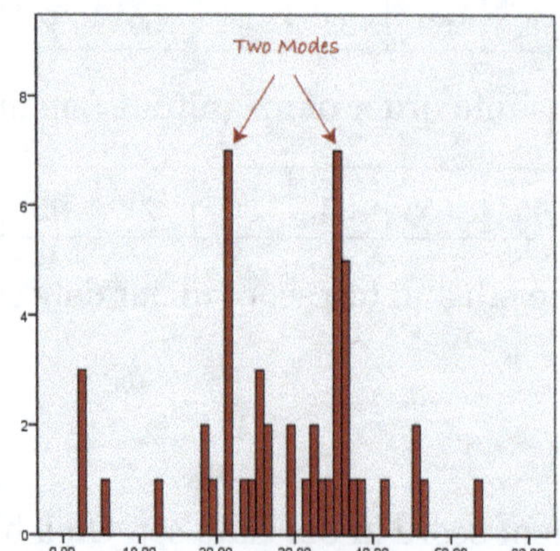

We are now stuck as to which mode best describes the central tendency of the data. This is particularly problematic when we have continuous data because we are more likely not to have any one value that is more frequent than the other. For example, consider measuring 30 peoples' weight (to the nearest 0.1 kg). How likely is it that we will find two or more people with exactly the same weight (e.g., 67.4 kg)? The answer, is probably very unlikely - many people might be close, but with such a small sample (30 people) and a large range of possible weights, you are unlikely to find two people with exactly the same weight; that is, to the nearest 0.1 kg. This is why the mode is very rarely used with continuous data.

Another problem with the mode is that it will not provide us with a very good measure of central tendency when the most common mark is far away from the rest of the data in the data set, as depicted in the diagram below:

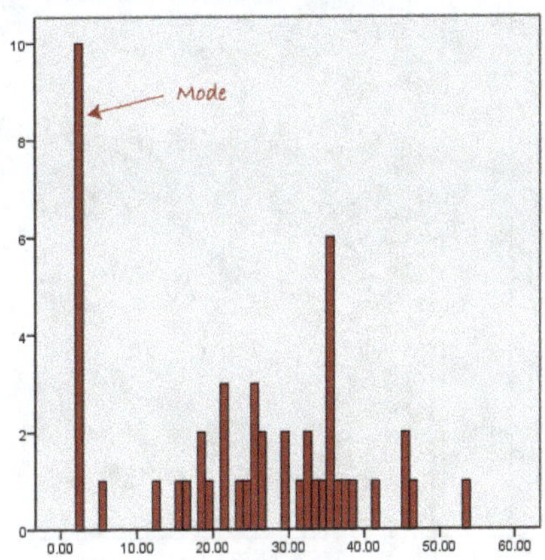

In the above diagram the mode has a value of 2. We can clearly see, however, that the mode is not representative of the data, which is mostly concentrated around the 20 to 30 value range. To use the mode to describe the central tendency of this data set would be misleading.

Skewed Distributions and the Mean and Median

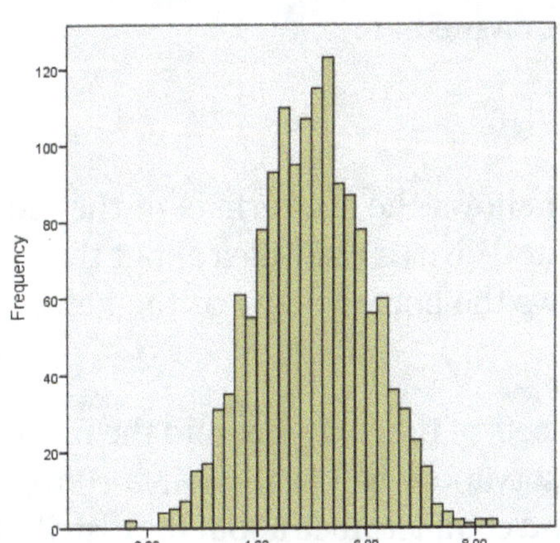

We often test whether our data is normally distributed because this is a common assumption underlying many statistical tests. An example of a normally distributed set of data is presented below:

When you have a normally distributed sample you can legitimately use both the mean or the median as your measure of central tendency. In fact, in any symmetrical distribution the mean, median and mode are equal. However, in this situation, the mean is widely preferred as the best measure of central tendency because it is the measure that includes all the values in the data set for its calculation, and any change in any of the scores will affect the value of the mean. This is not the case with the median or mode.

However, when our data is skewed, for example, as with the right-skewed data set below:

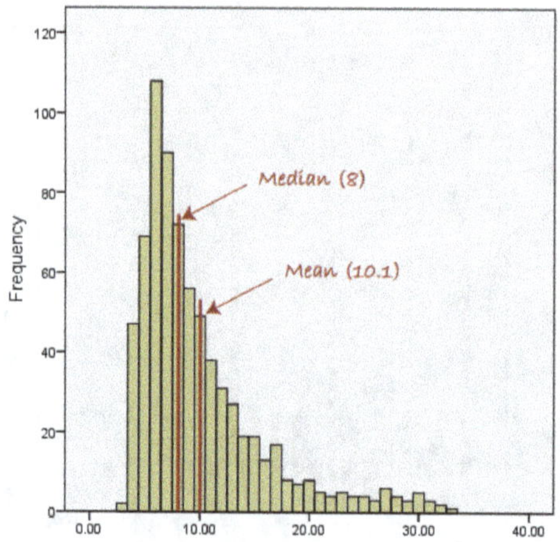

We find that the mean is being dragged in the direct of the skew. In these situations, the median is generally considered to be the best representative of the central location of the data. The more skewed the distribution, the greater the difference between the median and mean, and the greater emphasis should be placed on using the median as opposed to the mean. A classic example of the above right-skewed distribution is income (salary), where higher-earners provide a false representation of the typical income if expressed as a mean and not a median.

Measures of Dispersion

The measure of dispersion shows the scatterings of the data. It tells the variation of the data from one another and gives a clear idea about the distribution of the data. The measure of dispersion shows the homogeneity or the heterogeneity of the distribution of the observations.

Suppose you have four datasets of the same size and the mean is also same, say, m. In all the cases the sum of the observations will be the same. Here, the measure of central tendency is not giving a clear and complete idea about the distribution for the four given sets.

Can we get an idea about the distribution if we get to know about the dispersion of the observations from one another within and between the datasets? The main idea about the measure of dispersion is to get to know how the data are spread. It shows how much the data vary from their average value.

Characteristics of Measures of Dispersion

- A measure of dispersion should be rigidly defined
- It must be easy to calculate and understand
- Not affected much by the fluctuations of observations
- Based on all observations

Classification of Measures of Dispersion

The measure of dispersion is categorized as:

(i) An absolute measure of dispersion-

- The measures which express the scattering of observation in terms of distances i.e., range, quartile deviation.
- The measure which expresses the variations in terms of the average of deviations of observations like mean deviation and standard deviation.

(ii) A relative measure of dispersion-

We use a relative measure of dispersion for comparing distributions of two or more data set and for unit free comparison. They are the coefficient of range, the coefficient of mean deviation, the coefficient of quartile deviation, the coefficient of variation, and the coefficient of standard deviation.

Range

A range is the most common and easily understandable measure of dispersion. It is the difference between two extreme observations of the data set. If X_{max} and X_{min} are the two extreme observations then

$$\text{Range} = X_{max} - X_{min}$$

Merits of Range

- It is the simplest of the measure of dispersion
- Easy to calculate

- Easy to understand
- Independent of change of origin.

Demerits of Range

- It is based on two extreme observations. Hence, get affected by fluctuations
- A range is not a reliable measure of dispersion
- Dependent on change of scale.

Quartile Deviation

The quartiles divide a data set into quarters. The first quartile, (Q_1) is the middle number between the smallest number and the median of the data. The second quartile, (Q_2) is the median of the data set. The third quartile, (Q_3) is the middle number between the median and the largest number.

Quartile deviation or semi-inter-quartile deviation is,

$$Q = \tfrac{1}{2} \times (Q_3 - Q_1)$$

Merits of Quartile Deviation

- All the drawbacks of Range are overcome by quartile deviation
- It uses half of the data
- Independent of change of origin
- The best measure of dispersion for open-end classification

Demerits of Quartile Deviation

- It ignores 50% of the data
- Dependent on change of scale
- Not a reliable measure of dispersion

Mean Deviation

Mean deviation is the arithmetic mean of the absolute deviations of the observations from a measure of central tendency. If $x_1, x_2, \ldots, x_n$ are the set of observation, then the mean deviation of x about the average A (mean, median, or mode) is

$$\text{Mean deviation from average } A = \tfrac{1}{n}\left[\textstyle\sum_i |x_i - A|\right]$$

For a grouped frequency, it is calculated as:

Mean deviation from average $A = {}^1/N\,[\Sigma_i\ f_i\,|\,x_i - A\,|]$, $N = \Sigma f_i$

Here, x_i and f_i are respectively the mid value and the frequency of the ith class interval.

Merits of Mean Deviation

- Based on all observations
- It provides a minimum value when the deviations are taken from the median
- Independent of change of origin

Demerits of Mean Deviation

- Not easily understandable
- Its calculation is not easy and time-consuming
- Dependent on the change of scale
- Ignorance of negative sign creates artificiality and becomes useless for further mathematical treatment.

Standard Deviation

A standard deviation is the positive square root of the arithmetic mean of the squares of the deviations of the given values from their arithmetic mean. It is denoted by a Greek letter sigma, σ. It is also referred to as root mean square deviation. The standard deviation is given as,

$$\sigma = [(\Sigma_i (y_i - \bar{y})\ \bar{y}n]^{1/2} = [(\Sigma_i y_i^2 / n) - \bar{y}^2]^{1/2}$$

For a grouped frequency distribution, it is,

$$\sigma = [(\Sigma_i\ f_i\,(y_i - \bar{y})/N]^{1/2} = [(\Sigma_i f_i\ y_i^2/n) - \bar{y}^2]^{1/2}$$

The square of the standard deviation is the variance. It is also a measure of dispersion.

$$\sigma^2 = [(\Sigma_i (y_i - \bar{y}) / n]^{1/2} = [(\Sigma_i y_i^2/n)\ - \bar{y}^2]$$

For a grouped frequency distribution, it is,

$$\sigma^2 = [(\Sigma_i\ f_i\,(y_i\ - \bar{y})/N]^{1/2} = [(\Sigma_i\ f_i\ x_i^2/n) - \bar{y}^2].$$

If instead of a mean, we choose any other arbitrary number, say A, the standard deviation becomes the root mean deviation.

Variance of the Combined Series

If σ_1, σ_2 are two standard deviations of two series of sizes n_1 and n_2 with means $\bar{y}_1$ and $\bar{y}_2$. The variance of the two series of sizes n_1+n_2 is:

$$\sigma^2 = (1/n_1 + n_2) \div [n_1(\sigma_1^2 + d_1^2) + n_2(\sigma_2^2 + d_2^2)]$$

where, $d_1 = \bar{y}_1 - \bar{y}$, $d_2 = \bar{y}_2 - \bar{y}$, and $\bar{y} = (n_1\bar{y}_1 + n_2\bar{y}_2) \div (n_1 + n_2)$.

Merits of Standard Deviation

- Squaring the deviations overcomes the drawback of ignoring signs in mean deviations
- Suitable for further mathematical treatment
- Least affected by the fluctuation of the observations
- The standard deviation is zero if all the observations are constant
- Independent of change of origin.

Demerits of Standard Deviation

- Not easy to calculate
- Difficult to understand for a layman
- Dependent on the change of scale.

Coefficient of Dispersion

Whenever we want to compare the variability of the two series which differ widely in their averages. Also, when the unit of measurement is different. We need to calculate the coefficients of dispersion along with the measure of dispersion. The coefficients of dispersion (C.D.) based on different measures of dispersion are,

- Based on Range = $(X_{max} - X_{min})/(X_{max} + X_{min})$.
- C.D. based on quartile deviation = $(Q_3 - Q1)/(Q_3 + Q1)$.
- Based on mean deviation = Mean deviation/average from which it is calculated.
- For Standard deviation = S.D./Mean

Coefficient of Variation

100 times the coefficient of dispersion based on standard deviation is the coefficient of variation (C.V.).

$$C.V. = 100 \times (S.D./Mean) = (\sigma / \bar{y}) \times 100.$$

Solved example on measures of dispersion

Problem: Below is the table showing the values of the results for two companies A, and B.

	Company A	Company B
Number of employees	900	1000
Average daily wage	Rs. 250	Rs. 220
Variance in the distribution of Wages	100	144

1. Which of the company has a larger wage bill?
2. Calculate the coefficients of variations for both of the companies.
3. Calculate the average daily wage and the variance of the distribution of wages of all the employees in the firms A and B taken together.

Solution:

For Company A

No. of employees = n_1 = 900, and average daily wages = $\bar{y}_1$ = Rs. 250

We know, average daily wage = Total wages/Total number of employees

or, Total wages = Total employees × average daily wage = 900 × 250 = Rs. 225000 ... (i)

For Company B

No. of employees = n_2 = 1000, and average daily wages = $\bar{y}_2$ = Rs. 220

So, Total wages = Total employees × average daily wage = 1000 × 220 = Rs. 220000 ... (ii)

Comparing (i), and (ii), we see that Company A has a larger wage bill.

For Company A

Variance of distribution of wages $= \sigma_1^2 = 100$

C.V. of distribution of wages = 100 x standard deviation of distribution of wages/ average daily wages

$$\text{Or, } C.V._A = 100 \times \sqrt{100}/250 = 100 \times 10/250 = 4 \ldots (i)$$

For Company B

Variance of distribution of wages $= \sigma_2{}^2 = 144$

$$C.V._B = 100 \times \sqrt{144}/220 = 100 \times 12/220 = 5.45 \ldots (ii)$$

Comparing (i), and (ii), we see that Company B has greater variability.

For Company A and B Taken Together

The average daily wages for both the companies taken together,

$$\bar{y} = (n_1\bar{y}_1 + n_2\bar{y}_2)/(n_1 + n_2) = (900 \times 250 + 1000 \times 220) \div (900 + 1000)$$

$$= 445000/1900 = Rs.234.21$$

$$\text{The combined variance}, \sigma^2 = (1/n_1 + n_2) \div [n_1(\sigma_1{}^2 + d_1{}^2) + n_2(\sigma_2{}^2 + d_2{}^2)]$$

$$\text{Here, } d_1 = \bar{y}_1 - \bar{y} = 250 - 234.21 = 15.79, d_2 = \bar{y}_2 - \bar{y} = 220 - 234.21 = -14.21.$$

$$\text{Hence}, \sigma^2 = [900 \times (100 + 15.79^2) + 1000 \times (144 + -14.21^2)]/(900 + 1000)$$

$$\text{or}, \sigma^2 = (314391.69 + 345924.10)/1900 = 347.53.$$

Inferential Statistics

Inferential statistics is one of the two main branches of statistics.

Inferential statistics use a random sample of data taken from a population to describe and make inferences about the population. Inferential statistics are valuable when examination of each member of an entire population is not convenient or possible. For example, to measure the diameter of each nail that is manufactured in a mill is impractical. You can measure the diameters of a representative random sample of nails. You can use the information from the sample to make generalizations about the diameters of all of the nails.

Types of Inferential Statistics Tests

There are many tests in this field, of which some of the most important are mentioned below.

1. Linear Regression Analysis

In this test, a linear algorithm is used to understand the relationship between two variables from the data set. One of those variables is the dependent variable, while there can be one or more independent variables used. In simpler terms, we try to predict the value of the dependent variable based on the available values of the independent variables. This is usually represented by using a scatter plot, although we can also use other types of graphs too.

2. Analysis of Variance

This is another statistical method which is extremely popular in data science. It is used to test and analyse the differences between two or more means from the data set. The significant differences between the means are obtained, using this test.

3. Analysis of Co-variance

This is only a development on the Analysis of Variance method and involves the inclusion of a continuous co-variance in the calculations. A co-variate is an independent variable which is continuous, and are used as regression variables. This method is used extensively in statistical modelling, in order to study the differences present between the average values of dependent variables.

4. Statistical Significance (T-Test)

A relatively simple test in inferential statistics, this is used to compare the means of two groups and understand if they are different from each other. The order of difference, or how significant the differences are can be obtained from this.

5. Correlation Analysis

Another extremely useful test, this is used to understand the extent to which two variables are dependent on each other. The strength of any relationship, if they exist, between the two variables can be obtained from this. You will be able to understand whether the variables have a strong correlation or a weak one. The correlation can also be negative or positive, depending upon the variables. A negative correlation means that the value of one variable decreases while the value of the other increases and positive correlation means that the value both variables decrease or increase simultaneously.

Applications of Statistics

Mathematics

The formulas used in math are reliable, but to get more precision and exactness, statistics methods are important. In fact, it is called the branch of applied math. There are common techniques that both the fields have adopted from each other such as statistical methods, namely probability, dispersion, etc., used in math and mathematical concepts like integration and algebra are used in former.

Business

Business students must be aware of the importance of statistics in the field. There are times when a businessman has to make quick decisions, and this can be done by using its concepts which make the decision-making easy. He strategizes the marketing, finance, production, resource through it. What are the tastes and preferences of consumers? What should be the quality? What should be the target market? All these questions are answered using statistical tools.

Economics

There are so many concepts of economics that are completely dependent on statistics. All the data collected to find out the national income, employment, inflation, etc., are interpreted through it. In fact, theory of demand and supply, relationship between exports and imports are studied through this subject. The perfect example of this is census; the bureau uses its formulas for calculating a country's population.

Country's Administration

Many national policies are decided using statistical methods, and administrative decisions are taken based on its data. Statistics provides most accurate data which helps government to make budgets and estimate expenditures and revenues. It is also used to revise the pay scale of employees in case cost of living is rising.

Astronomy

When scientists measured the distance between sun and earth, or moon and earth, they did not use any measurement scale or ruler for that. It was these statistical methods that helped them to find out the best answers and estimates that are possible. It is difficult to measure the mass, size, distance, density of objects in the universe without any error, but statistics formulas do this with the best probability.

Banking

When someone deposits his money in banks, the idea is that he will not withdraw the amount in the near future. So, banks lend this money to other customers to earn profit in the form of interest. They use statistical approach for this service. They compare the number of people making deposits against the number of people requesting loans and at the same time ascertaining the estimated day for the claim.

Accounting and Auditing

Although accounting needs exactness in calculating the profit and loss of the business, certain decisions can be taken according to approximation which is done through sta-

tistics. For example, sampling may be used to find out the current trends in the market as it does not require any precision.

Sociology

Sociology is one of the social sciences aiming to discover the basic structure of human society, to identify the main forces that hold groups together or weaken them and to learn the conditions that transform social life. It highlights and illuminates aspects of social life that otherwise might be only obscurely recognized and understood. The sociologist may be called upon for help with a special problem such as social conflict, urban plight or the war on poverty or crimes. His practical contribution lies in the ability to clarify the underlaying nature of social problems to estimate more exactly their dimensions and to identify aspects that seem most amenable to remedy with the knowledge and skills at hand.

He naturally lands in sociological research which is the purposeful effort to learn more about society than one can in the ordinary course of living. Keeping in view of the problem he sets forth his objectives collects materials or data and uses statistical techniques and the knowledge and theory already established on similar topics to achieve his objectives. So statistical data and statistical methods are quite indispensable for sociological research studies. There is a growing emphasis recently on social survey methods or research methodology in all faculties of arts.

Sociologists seek the help of statistical tools to study cultural change in the society, family pattern,prostitution,crime,marriage system etc.They also study statistically the relation between prostitution and poverty, crime and poverty,drunkness and crime, illiteracy and crime etc.Thus statistics is of immense use in various sociological studies.

Government

The functions of a government are more varied and complex. Various depts in the state are required to collect and record statistical data in a systematic manner for an effective administration. Data pertaining to various fields namely population, natural resources, production both agricultural and industrial,finance,trade,exports and imports, prices, labor, transport and communication, health, education,defence ,crimes etc are the most fundamental requirements of the state for its administration. It is only on this basis of such data; the government decides on the priority areas, gives more attention to them through target oriented programmes and studies the impact of the programmes for its future guidelines.

Planning

Modern age is an age of planning and statistics are indispensable for planning. According to Tippett planning greater or lesser degree according to the government in power is the order of the day and without statistics, planning is inconceivable. Based only on a correct assessment of various resources both human and material of the country proper planning can be made. A study of data relating to population, agriculture, industry, prices, employment, health, education enables the planners to fix up time-bound targets on the social and economic fronts evaluation of such economic and social programmes at different stages by means of related data gathered continuously and systematically is also done to decide whether the programmes are on towards the goal or targets set.

References

- What-is-probability-in-math: study.com, Retrieved 11 March, 2019
- What-is-conditional-probability: datasciencecentral.com, Retrieved 21 July, 2019
- Probability, statistics-handling-data, maths: revisionmaths.com, Retrieved 28 June, 2019
- Probability-theory, science: britannica.com, Retrieved 23 April, 2019
- Bayes-theorem: thoughtco.com, Retrieved 13 May, 2019
- Statistics, maths: byjus.com, Retrieved 19 March, 2019
- Measures-central-tendency-mean-mode-median: laerd.com, Retrieved 9 January, 2019
- Measure-of-dispersion, measures-of-central-tendency-and-dispersion, business-mathematics-and-statistics: toppr.com, Retrieved 12 February, 2019
- What-are-inferential-statistics, basic-statistics: minitab.com, Retrieved 22 April, 2019
- Inferential-statistics-definition-and-types: acadgild.com, Retrieved 28 January, 2019
- Application-of-statistics: assignmentdesk.co.uk, Retrieved 20 August, 2019
- Applications-of-statistics, research-methods&statistics: sociologyguide.com, Retrieved 10 February, 2019

Permissions

All chapters in this book are published with permission under the Creative Commons Attribution Share Alike License or equivalent. Every chapter published in this book has been scrutinized by our experts. Their significance has been extensively debated. The topics covered herein carry significant information for a comprehensive understanding. They may even be implemented as practical applications or may be referred to as a beginning point for further studies.

We would like to thank the editorial team for lending their expertise to make the book truly unique. They have played a crucial role in the development of this book. Without their invaluable contributions this book wouldn't have been possible. They have made vital efforts to compile up to date information on the varied aspects of this subject to make this book a valuable addition to the collection of many professionals and students.

This book was conceptualized with the vision of imparting up-to-date and integrated information in this field. To ensure the same, a matchless editorial board was set up. Every individual on the board went through rigorous rounds of assessment to prove their worth. After which they invested a large part of their time researching and compiling the most relevant data for our readers.

The editorial board has been involved in producing this book since its inception. They have spent rigorous hours researching and exploring the diverse topics which have resulted in the successful publishing of this book. They have passed on their knowledge of decades through this book. To expedite this challenging task, the publisher supported the team at every step. A small team of assistant editors was also appointed to further simplify the editing procedure and attain best results for the readers.

Apart from the editorial board, the designing team has also invested a significant amount of their time in understanding the subject and creating the most relevant covers. They scrutinized every image to scout for the most suitable representation of the subject and create an appropriate cover for the book.

The publishing team has been an ardent support to the editorial, designing and production team. Their endless efforts to recruit the best for this project, has resulted in the accomplishment of this book. They are a veteran in the field of academics and their pool of knowledge is as vast as their experience in printing. Their expertise and guidance has proved useful at every step. Their uncompromising quality standards have made this book an exceptional effort. Their encouragement from time to time has been an inspiration for everyone.

The publisher and the editorial board hope that this book will prove to be a valuable piece of knowledge for students, practitioners and scholars across the globe.

Index

www.ingramcontent.com/pod-product-compliance
Lightning Source LLC
LaVergne TN
LVHW080611210926
848485LV00003B/138